Science for Environmental Protection
THE ROAD AHEAD

Committee on Science for EPA's Future

Board on Environmental Studies and Toxicology

Division on Earth and Life Studies

National Research Council

NATIONAL RESEARCH COUNCIL
OF THE NATIONAL ACADEMIES

THE NATIONAL ACADEMIES PRESS
Washington, D.C.
www.nap.edu

THE NATIONAL ACADEMIES PRESS   500 Fifth Street, NW   Washington, DC 20001

NOTICE: The project that is the subject of this report was approved by the Governing Board of the National Research Council, whose members are drawn from the councils of the National Academy of Sciences, the National Academy of Engineering, and the Institute of Medicine. The members of the committee responsible for the report were chosen for their special competences and with regard for appropriate balance.

This project was supported by contract number EP-C-09-003, TO#: 13 between the National Academy of Sciences and the US Environmental Protection Agency. Any opinions, findings, conclusions, or recommendations expressed in this publication are those of the authors and do not necessarily reflect the view of the organizations or agencies that provided support for this project.

International Standard Book Number-13: 978-0-309-26489-1
International Standard Book Number-10: 0-309-26489-8
Library of Congress Control Number: 2012951535

Additional copies of this report are available for sale from the National Academies Press, 500 Fifth Street, NW, Keck 360, Washington, DC 20001; (800) 624-6242 or (202) 334-3313; http://www.nap.edu/.

Copyright 2012 by the National Academy of Sciences. All rights reserved.

Printed in the United States of America

THE NATIONAL ACADEMIES
Advisers to the Nation on Science, Engineering, and Medicine

The **National Academy of Sciences** is a private, nonprofit, self-perpetuating society of distinguished scholars engaged in scientific and engineering research, dedicated to the furtherance of science and technology and to their use for the general welfare. Upon the authority of the charter granted to it by the Congress in 1863, the Academy has a mandate that requires it to advise the federal government on scientific and technical matters. Dr. Ralph J. Cicerone is president of the National Academy of Sciences.

The **National Academy of Engineering** was established in 1964, under the charter of the National Academy of Sciences, as a parallel organization of outstanding engineers. It is autonomous in its administration and in the selection of its members, sharing with the National Academy of Sciences the responsibility for advising the federal government. The National Academy of Engineering also sponsors engineering programs aimed at meeting national needs, encourages education and research, and recognizes the superior achievements of engineers. Dr. Charles M. Vest is president of the National Academy of Engineering.

The **Institute of Medicine** was established in 1970 by the National Academy of Sciences to secure the services of eminent members of appropriate professions in the examination of policy matters pertaining to the health of the public. The Institute acts under the responsibility given to the National Academy of Sciences by its congressional charter to be an adviser to the federal government and, upon its own initiative, to identify issues of medical care, research, and education. Dr. Harvey V. Fineberg is president of the Institute of Medicine.

The **National Research Council** was organized by the National Academy of Sciences in 1916 to associate the broad community of science and technology with the Academy's purposes of furthering knowledge and advising the federal government. Functioning in accordance with general policies determined by the Academy, the Council has become the principal operating agency of both the National Academy of Sciences and the National Academy of Engineering in providing services to the government, the public, and the scientific and engineering communities. The Council is administered jointly by both Academies and the Institute of Medicine. Dr. Ralph J. Cicerone and Dr. Charles M. Vest are chair and vice chair, respectively, of the National Research Council.

www.national-academies.org

COMMITTEE ON SCIENCE FOR EPA'S FUTURE

Members

JERALD L. SCHNOOR (*Chair*), University of Iowa, Iowa City
TINA BAHADORI, American Chemistry Council, Washington, DC
    (resigned March 23, 2012)
ERIC J. BECKMAN, University of Pittsburgh, Pittsburgh, PA
THOMAS A. BURKE, Johns Hopkins Bloomberg School of Public Health,
    Baltimore, MD
FRANK W. DAVIS, University of California, Santa Barbara
DAVID L. EATON, University of Washington, Seattle
PAUL GILMAN, Covanta Energy, Fairfield, NJ
DANIEL S. GREENBAUM, Health Effects Institute, Boston, MA
STEVEN P. HAMBURG, Environmental Defense Fund, Boston, MA
JAMES E. HUTCHISON, University of Oregon, Eugene
JONATHAN I. LEVY, Boston University, Boston, MA
DAVID E. LIDDLE, U.S. Venture Partners, Menlo Park, CA
JANA B. MILFORD, University of Colorado, Boulder
M. GRANGER MORGAN, Carnegie Mellon University, Pittsburgh, PA
ANA NAVAS-ACIEN, Johns Hopkins Bloomberg School of Public Health,
    Baltimore, MD
GORDON H. ORIANS, University of Washington (retired), Lake Forest Park, WA
JOAN B. ROSE, Michigan State University, East Lansing
JAMES S. SHORTLE, Pennsylvania State University, University Park
JOEL A. TICKNER, University of Massachusetts, Lowell
ANTHONY D. WILLIAMS, Anthony D. Williams Consulting, Toronto,
    Ontario, Canada
YILIANG ZHU, University of South Florida, Tampa

Staff

HEIDI MURRAY-SMITH, Project Director
JAMES REISA, Director, Board on Environmental Studies and Toxicology
DAVID POLICANSKY, Scholar
KERI STOEVER, Research Associate
NORMAN GROSSBLATT, Senior Editor
MIRSADA KARALIC-LONCAREVIC, Manager, Technical Information Center
RADIAH ROSE, Manager, Editorial Projects
CRAIG PHILIP, Senior Program Assistant

Sponsor

US ENVIRONMENTAL PROTECTION AGENCY

BOARD ON ENVIRONMENTAL STUDIES AND TOXICOLOGY[1]

Members

ROGENE F. HENDERSON (*Chair*), Lovelace Respiratory Research Institute, Albuquerque, NM
PRAVEEN AMAR, Clean Air Task Force, Boston, MA
MICHAEL J. BRADLEY, M.J. Bradley & Associates, Concord, MA
JONATHAN Z. CANNON, University of Virginia, Charlottesville
GAIL CHARNLEY, HealthRisk Strategies, Washington, DC
FRANK W. DAVIS, University of California, Santa Barbara
RICHARD A. DENISON, Environmental Defense Fund, Washington, DC
CHARLES T. DRISCOLL, JR., Syracuse University, New York
H. CHRISTOPHER FREY, North Carolina State University, Raleigh
RICHARD M. GOLD, Holland & Knight, LLP, Washington, DC
LYNN R. GOLDMAN, George Washington University, Washington, DC
LINDA E. GREER, Natural Resources Defense Council, Washington, DC
WILLIAM E. HALPERIN, University of Medicine and Dentistry of New Jersey, Newark
PHILIP K. HOPKE, Clarkson University, Potsdam, NY
HOWARD HU, University of Michigan, Ann Arbor
SAMUEL KACEW, University of Ottawa, Ontario
ROGER E. KASPERSON, Clark University, Worcester, MA
THOMAS E. MCKONE, University of California, Berkeley
TERRY L. MEDLEY, E.I. du Pont de Nemours & Company, Wilmington, DE
JANA MILFORD, University of Colorado at Boulder, Boulder
FRANK O'DONNELL, Clean Air Watch, Washington, DC
RICHARD L. POIROT, Vermont Department of Environmental Conservation, Waterbury
KATHRYN G. SESSIONS, Health and Environmental Funders Network, Bethesda, MD
JOYCE S. TSUJI, Exponent Environmental Group, Bellevue, WA

Senior Staff

JAMES J. REISA, Director
DAVID J. POLICANSKY, Scholar
RAYMOND A. WASSEL, Senior Program Officer for Environmental Studies
ELLEN K. MANTUS, Senior Program Officer for Risk Analysis
SUSAN N.J. MARTEL, Senior Program Officer for Toxicology
EILEEN N. ABT, Senior Program Officer
MIRSADA KARALIC-LONCAREVIC, Manager, Technical Information Center
RADIAH ROSE, Manager, Editorial Projects

---

[1] This study was planned, overseen, and supported by the Board on Environmental Studies and Toxicology.

OTHER REPORTS OF THE
BOARD ON ENVIRONMENTAL STUDIES AND TOXICOLOGY

Exposure Science in the 21st Century: A Vision and A Strategy (2012)
A Research Strategy for Environmental, Health, and Safety Aspects of Engineered Nanomaterials (2012)
Macondo Well–Deepwater Horizon Blowout: Lessons for Improving Offshore Drilling Safety (2012)
Feasibility of Using Mycoherbicides for Controlling Illicit Drug Crops (2011)
Improving Health in the United States: The Role of Health Impact Assessment (2011)
A Risk-Characterization Framework for Decision-Making at the Food and Drug Administration (2011)
Review of the Environmental Protection Agency's Draft IRIS Assessment of Formaldehyde (2011)
Toxicity-Pathway-Based Risk Assessment: Preparing for Paradigm Change (2010)
The Use of Title 42 Authority at the U.S. Environmental Protection Agency (2010)
Review of the Environmental Protection Agency's Draft IRIS Assessment of Tetrachloroethylene (2010)
Hidden Costs of Energy: Unpriced Consequences of Energy Production and Use (2009)
Contaminated Water Supplies at Camp Lejeune—Assessing Potential Health Effects (2009)
Review of the Federal Strategy for Nanotechnology-Related Environmental, Health, and Safety Research (2009)
Science and Decisions: Advancing Risk Assessment (2009)
Phthalates and Cumulative Risk Assessment: The Tasks Ahead (2008)
Estimating Mortality Risk Reduction and Economic Benefits from Controlling Ozone Air Pollution (2008)
Respiratory Diseases Research at NIOSH (2008)
Evaluating Research Efficiency in the U.S. Environmental Protection Agency (2008)
Hydrology, Ecology, and Fishes of the Klamath River Basin (2008)
Applications of Toxicogenomic Technologies to Predictive Toxicology and Risk Assessment (2007)
Models in Environmental Regulatory Decision Making (2007)
Toxicity Testing in the Twenty-first Century: A Vision and a Strategy (2007)
Sediment Dredging at Superfund Megasites: Assessing the Effectiveness (2007)
Environmental Impacts of Wind-Energy Projects (2007)
Scientific Review of the Proposed Risk Assessment Bulletin from the Office of Management and Budget (2007)
Assessing the Human Health Risks of Trichloroethylene: Key Scientific Issues (2006)
New Source Review for Stationary Sources of Air Pollution (2006)
Human Biomonitoring for Environmental Chemicals (2006)
Health Risks from Dioxin and Related Compounds: Evaluation of the EPA Reassessment (2006)

Fluoride in Drinking Water: A Scientific Review of EPA's Standards (2006)
State and Federal Standards for Mobile-Source Emissions (2006)
Superfund and Mining Megasites—Lessons from the Coeur d'Alene River
    Basin (2005)
Health Implications of Perchlorate Ingestion (2005)
Air Quality Management in the United States (2004)
Endangered and Threatened Species of the Platte River (2004)
Atlantic Salmon in Maine (2004)
Endangered and Threatened Fishes in the Klamath River Basin (2004)
Cumulative Environmental Effects of Alaska North Slope Oil and Gas
    Development (2003)
Estimating the Public Health Benefits of Proposed Air Pollution Regulations (2002)
Biosolids Applied to Land: Advancing Standards and Practices (2002)
The Airliner Cabin Environment and Health of Passengers and Crew (2002)
Arsenic in Drinking Water: 2001 Update (2001)
Evaluating Vehicle Emissions Inspection and Maintenance Programs (2001)
Compensating for Wetland Losses Under the Clean Water Act (2001)
A Risk-Management Strategy for PCB-Contaminated Sediments (2001)
Acute Exposure Guideline Levels for Selected Airborne Chemicals (twelve
    volumes, 2000-2012)
Toxicological Effects of Methylmercury (2000)
Strengthening Science at the U.S. Environmental Protection Agency (2000)
Scientific Frontiers in Developmental Toxicology and Risk Assessment (2000)
Ecological Indicators for the Nation (2000)
Waste Incineration and Public Health (2000)
Hormonally Active Agents in the Environment (1999)
Research Priorities for Airborne Particulate Matter (four volumes, 1998-2004)
The National Research Council's Committee on Toxicology: The First 50
    Years (1997)
Carcinogens and Anticarcinogens in the Human Diet (1996)
Upstream: Salmon and Society in the Pacific Northwest (1996)
Science and the Endangered Species Act (1995)
Wetlands: Characteristics and Boundaries (1995)
Biologic Markers (five volumes, 1989-1995)
Science and Judgment in Risk Assessment (1994)
Pesticides in the Diets of Infants and Children (1993)
Dolphins and the Tuna Industry (1992)
Science and the National Parks (1992)
Human Exposure Assessment for Airborne Pollutants (1991)
Rethinking the Ozone Problem in Urban and Regional Air Pollution (1991)
Decline of the Sea Turtles (1990)

Copies of these reports may be ordered from the National Academies Press
(800) 624-6242 or (202) 334-3313
www.nap.edu

Preface

The important thing in science is not so much to obtain new facts as to discover new ways of thinking about them. —William Lawrence Bragg

Environmental protection in the 21st century requires a new way of thinking about pollution and its drivers, scale, effects, and solutions. In the United States, the Environmental Protection Agency (EPA) has the mission of protecting human health and the environment. EPA helps to identify emerging and future problems, it assesses the fate and effects of pollutants, and it researches methods for prevention, intervention, and remediation. Science at EPA should be relevant to the agency's mission, it should be of high quality and high priority, and it should be continuously reviewed by peer scientists, engineers, and social scientists.

With a 42-year history, EPA finds itself in the second decade of the new millennium with different challenges and variable public support for its mission to protect human health and the environment. It has successfully controlled pollution and improved public health and welfare since it was formed in 1970. Its success has stemmed largely from the establishment and enforcement of its regulatory programs under the Safe Drinking Water Act; the Clean Air Act; the Federal Insecticide, Fungicide, and Rodenticide Act; the Comprehensive Environmental Response, Compensation, and Liability Act; the Toxic Substances Control Act; and other statutes. That success has been informed by research within the agency and outside the agency by academe, nonprofit organizations, industry consultants, federal agencies, and other partnering agencies and institutions.

Many of today's problems present challenges of great scope, spatial scale, and complexity. Some pose a new suite of emerging environmental threats, while others persist and have yet to be completely solved. Examples of today's environmental problems include the deterioration of air quality due to a warmer, moister climate; effects of the energy production required to fuel a modern, growing economy; hypoxia, harmful algal blooms, and eutrophication from agricultural runoff and nutrient pollution; overload of urban stormwater and bypass of raw sewage exacerbated by sprawl and storm severity; and loss of species due to land use and climate change. Those are problems of the 21st century, and addressing them will require the best available science and technology possible in a resource-constrained world.

In 2011, EPA asked the National Research Council to assess independently the overall capabilities of the agency to develop, obtain, and use the best available scientific and technologic information and tools to meet persistent, emerging, and future mission challenges and opportunities. In response, the National Research Council convened the Committee on Science for EPA's Future, which prepared this report. The committee brings together a wide array of expertise to address major changes in the biophysical and societal environment, including risk assessment and management, computational techniques and bioinformatics, data mining and assimilation, crowd sourcing, benefit–cost analyses of environmental regulations, developments in public health, and organizational collaborations within EPA and beyond. The committee also called on numerous people from within EPA and experts who collaborate with EPA for their perspectives and insight on science for EPA's future. It assessed the major drivers of environmental change and tried to describe characteristics of the challenges of coming decades, discussed emerging tools and technologies that can be brought to bear on those challenges, and formulated some principles for how to build environmental protection in the 21st century while enhancing EPA's leadership and capacity.

This present report has been reviewed in draft form by persons chosen for their diverse perspectives and technical expertise in accordance with procedures approved by the National Research Council Report Review Committee. The purpose of the independent review is to provide candid and critical comments that will assist the institution in making its published report as sound as possible and to ensure that the report meets institutional standards of objectivity, evidence, and responsiveness to the study charge. The review comments and draft manuscript remain confidential to protect the integrity of the deliberative process. We thank the following for their review of the report: John C. Bailar, III, The University of Chicago; Ann Bostrom, University of Washington; Charles M. Auer, Charles Auer & Associates, LLC; John P. Connolly, Anchor QEA, LLC; John Crittenden, Georgia Institute of Technology; Jerome J. Cura, The Science Collaborative; Bernard D. Goldstein, University of Pittsburgh Graduate School of Public Health; Chao-Jun Li, McGill University; David L. Macintosh, Environmental Health & Engineering, Inc.; Denise L. Mauzerall, Princeton University; Stephen Polasky, University of Minnesota; Joseph P. Rodricks, ENVIRON; Pamela Shubat, Minnesota Department of Health; Ponisseril Somasundaran, Columbia University; and Mark J. Utell, University of Rochester School of Medicine and Dentistry.

Although the reviewers listed above have provided many constructive comments and suggestions, they were not asked to endorse the conclusions or recommendations, nor did they see the final draft of the report before its release. The review of the report was overseen by the review coordinator, Edwin H. Clark II, Earth Policy Institute, and the review monitor, Mike Kavanaugh, Geo-

Preface

syntec Consultants. Appointed by the National Research Council, they were responsible for making certain that an independent examination of the report was carried out in accordance with institutional procedures and that all review comments were carefully considered. Responsibility for the final content of the report rests entirely with the committee and the institution.

The committee gratefully acknowledges Paul Anastas, Al McGartland, Iris Goodman, Kristen Keteles, Jeff Morris, Peter Preuss, and Kevin Teichman, of EPA, and David Miller, of the National Institute of Environmental Health Sciences, for making presentations to the committee.

The committee is also grateful for the assistance of the National Research Council staff in preparing this report. Staff members who contributed to the effort are Heidi Murray-Smith, project director; James Reisa, director of the Board on Environmental Studies and Toxicology; David Policansky, scholar; Keri Stoever, research associate; Norman Grossblatt, senior editor; Mirsada Karalic-Loncarevic, manager of the Technical Information Center; Radiah Rose, manager of editorial projects; and Craig Philip, senior program assistant.

I especially thank the members of the committee for their efforts throughout the development of this report.

> Jerald L. Schnoor, *Chair*
> Committee on Science for EPA's Future

Contents

SUMMARY **3**

1 INTRODUCTION **15**
   The Changing Nature of Environmental Problems, 16
   Science and Engineering at the US Environmental Protection Agency, 18
   The Committee's Task, 22
   Organization of the Report, 23
   References, 24

2 CHALLENGES OF THE 21st CENTURY **27**
   Major Factors Leading to Environmental Change, 28
   Environmental and Human Health Challenges, 32
   Summary, 46
   References, 48

**3 USING EMERGING SCIENCE AND TECHNOLOGIES
    TO ADDRESS PERSISTENT AND FUTURE
    ENVIRONMENTAL CHALLENGES** **54**
   A Simple Paradigm for Data-Driven, Science-Informed
     Decisions in the Environmental Protection Agency, 54
   Tools and Technologies to Address Challenges Related
     to Chemical Exposures, Human Health, and the Environment, 57
   Tools and Technologies to Address Challenges Related To
     Air Pollution and Climate Change, 69
   Tools and Technologies to Address Challenges Related To
     Water Quality, 73
   Tools and Technologies to Address Challenges Related To
     Shifting Spatial and Temporal Scales, 80
   Using New Science to Drive Safer Technologies and Products, 88
   Summary, 94
   References, 95

xiii

4	**BUILDING SCIENCE FOR ENVIRONMENTAL PROTECTION IN THE 21st CENTURY** 107
	Embracing Systems Thinking for Producing and Applying Science for Decisions: A 21st Century Framework for Science to Inform Decisions, 109
	Staying at the Leading Edge of Science, 110
	Enhanced Tools and Skills for Applying Systems Thinking To Inform Decisions, 132
	Synthesis and Evaluation for Decisions, 144
	Overarching Recommendation, 151
	References, 153
5	**ENHANCED SCIENTIFIC LEADERSHIP AND CAPACITY IN THE US ENVIRONMENTAL PROTECTION AGENCY** 161
	Enhanced Agency-Wide Science Leadership in the US Environmental Protection Agency, 162
	Realignment of the Office of Research and Development, 166
	Coordination of Science Efforts in the US Environmental Protection Agency, 168
	Strengthening Science Capacity, 170
	Integrity, Ethics, And Transparency in the US Environmental Protection Agency's Production and Use of Scientific Information, 178
	Strengthening Science in a Time of Tight Budgets, 182
	Summary, 182
	References, 183
6	**FINDINGS AND RECOMMENDATIONS**.................................. 187
	Systems Thinking, 189
	Enhanced Science Leadership, 192
	Strengthening Capacity, 194
	Science, Tools, and Technologies to Address Current and Future Challenges, 199
	Improved Management and Use of Large Datasets, 200
	Innovation, 201
	Strengthening Science in a Time of Tight Budgets, 203

APPENDIXES

A:	**STATEMENT OF TASK OF THE COMMITTEE ON SCIENCE FOR EPA'S FUTURE** 204
B:	**BIOGRAPHICAL INFORMATION ON THE COMMITTEE ON SCIENCE FOR EPA'S FUTURE** 206

Contents

C: THE RAPIDLY EXPANDING FIELD OF
   "–OMICS" TECHNOLOGIES 215

D: SCIENTIFIC COMPUTING, INFORMATION
   TECHNOLOGY, AND INFORMATICS 225

BOXES, FIGURES, AND TABLES

BOXES

2-1 Environmental Protection Agency Involvement in Climate Change, 39
3-1 Engaging the Public to Gather Information, 84
3-2 Example of Private Industry's Influence on the Supply Chain without Regulatory Mandates, 91
4-1 Principles to Guide the Development of Indicators, 124
4-2 Putting It All Together: The Case of Hydraulic Fracturing, 133
4-3 The Need for and Challenges of Life-Cycle Assessment: The Biofuels Case, 136
4-4 Example of a Solutions-Oriented Approach: Reducing Trichloroethylene Use in Massachusetts, 148
C-1 Comparison of Sanger and Next-Generation Sequencing (NGS), 217

FIGURES

S-1 Framework for enhanced science for environmental protection, 9
1-1 The Millennium Ecosystem Assessment conceptual framework, 18
1-2 Gross trends in drivers and aggregate emissions since 1980 in the United States, 20
2-1 Sources of phosphorus and nitrogen in the Gulf of Mexico and Chesapeake Bay, 44
2-2 Narragansett Bay nitrogen loading from 1850 to 2015 under several different scenarios, 45
3-1 The iterative process of science-informed environmental decision-making and policy, 56
3-2 Characterizing the exposome, 61
3-3 Schematic of an instrumented watershed in an observatory of the national network, 75
4-1 The iterative process of science-informed environmental decision-making and policy, 111
4-2 A framework for sustainable decisions at the US Environmental Protection Agency, 146
6-1 Framework for enhanced science for environmental protection, 190

TABLES

1-1 Change in Conventional Air Pollutant Emissions Over the Last 3 Decades, 20

2-1 Some Contrasts between the Clean Water Act and the Safe Drinking Water Act, 40
2-2 Large Regional Water Programs in the US Environmental Protection Agency, 47
3-2 Metagenomic Characterization of Pathogens and Microbial Populations in Biosolids, Wastewater, Rivers, and Lakes, 77
5-1 Former and Realigned Structures of EPA's Office of Research and Development, 167

Science for Environmental Protection

THE ROAD AHEAD

Summary

The stated mission of the US Environmental Protection Agency (EPA) is to protect human health and the environment. Since its formation in 1970, EPA has had a leadership role in developing many fields of environmental science and engineering. From ecology to health sciences and environmental engineering to analytic chemistry, EPA has performed, stimulated, and supported research; developed environmental education programs; supported regional science initiatives; supported safer technologies; and enhanced the scientific basis of informed decision-making. Science has always been an integral part of EPA's mission and is essential for providing the best-quality foundation of agency decisions. Today the agency's science is increasingly in the public eye, federal budgets are decreasing, and job creation and innovation are key national priorities.

In anticipation of future environmental science and engineering challenges and technologic advances, EPA asked the National Research Council (NRC) to assess the overall capabilities of the agency to develop, obtain, and use the best available scientific and technologic information and tools to meet persistent, emerging, and future mission challenges and opportunities. The NRC was also asked to identify and assess transitional options to strengthen the agency's capability to pursue and use scientific information and tools. In response, the NRC convened the Committee on Science for EPA's Future, which prepared the present report.

ENVIRONMENTAL CHALLENGES AND TOOLS TO ADDRESS THEM

The committee's report highlights a few persistent and emerging environmental challenges and tools and technologies to address them. Although the topics discussed in the report are only illustrative, the report provides specific examples and gives context to the committee's discussion of a broader framework for building science for environmental protection in the 21st century. Having assessed EPA's current activities, the committee notes that EPA is well equipped to take advantage of many scientific and technologic advances and that, in fact, its scientists and engineers are leaders in some fields.

Current and Persistent Environmental Challenges

There has been substantial progress over the last few decades in lessening many of the obvious environmental problems, such as black smoke coming from smokestacks, stench arising from rivers, and fish kills in US lakes. But the challenges associated with environmental protection today are complex, affected by many interacting factors, and no less daunting. They are on various spatial scales, may unfold over long temporal scales, and may have global implications. The problems are sometimes called "wicked problems", and are often characterized by being difficult to define, unstable, and socially complex; having no clear solution or end point; and extending beyond the understanding of one discipline or the responsibility of one organization. Although the committee cannot predict with certainty what new environmental problems EPA will face in the next 10 years or more, it can identify some of the common drivers and common characteristics of problems that are likely to occur. Some key features of persistent and future environmental challenges are complex feedback loops; the need to understand the effects of low-level exposures to numerous stressors as opposed to high-level exposures to individual stressors; the need to understand social, economic, and environmental drivers; and the need for systems thinking to devise optimal solutions.

The following are a few examples of persistent and emerging environmental challenges that pertain to EPA and its mission.

Chemical Exposures, Human Health, and the Environment. New chemicals continue to be created and enter the environment. Understanding what chemicals are in the environment, concentrations at which people are being exposed, pathways through which they are being exposed, and how different chemicals and stressors interact with one another encompasses some of the persistent challenges that EPA faces. Another challenge is to continue to elucidate the many factors that can modify the health effects of exposure to chemicals and other stressors. The chemical, biologic, and physical characteristics of an agent, the genetic and behavioral attributes of a host, and the physical and social characteristics of the environment are all influential.

Air Pollution and Climate Change. Emissions of major air pollutants were dramatically reduced from 1990 to 2010. Much of that success resulted from the establishment and enforcement of the Clean Air Act. Despite substantial progress, the agency's efforts to improve air quality continue to have high priority because the economic costs that air pollution imposes on society remain high. The Clean Air Act and other statutory mandates give rise to the need for improved scientific and technical information on health exposures and effects, on ecologic exposures and effects, on ambient and emission monitoring techniques, on atmospheric chemistry and physics, and on pollution-prevention and emission-control methods for hundreds of pollutants present in both indoor and outdoor environments. EPA also faces the critical challenge of helping to find efficient and effective approaches to mitigating climate change and improving understanding of how to adapt environmental management in the face of climate change.

Water Quality. The availability of clean water is essential for human consumption, personal hygiene, agriculture, business practices, recreation, and other activities. National water-quality policy has been driven primarily by the Clean Water Act and the Safe Drinking Water Act. With increasing demands on freshwater supplies, particularly in the more arid regions of the western United States, the challenges of providing freshwater are prominent today and will probably continue to be a concern in the future, especially as climate change alters water supply. Furthermore, water-quality challenges remain pressing, including the need to monitor and understand the transport and fate of contaminants, the need to maintain and update aging water-treatment infrastructure, and the need to address the persistent problem of nutrient pollution.

As progress has been made in solving local problems and as more has been learned about the health and environmental consequences of chronic low-level exposures to diverse and disperse physical and chemical stressors, environmental science and engineering has begun to focus on impacts over wider geographic areas. The spatial and temporal scales required to understand emerging environmental issues vary widely, and their range is widening as more is learned about the systems and feedback loops underlying the observed phenomena. These large-scale problems require improved understanding of the fate and transport of contaminants on international and global scales and of options for coordinated solutions. Long-term monitoring is also needed to identify and track changes and problems that develop slowly.

Developing Tools and Technologies to Address Environmental Challenges

Supporting the development of leading-edge scientific methods, tools, and technologies is critical for understanding environmental changes and their effects on human health and for identifying solutions. In addition, addressing the challenges of the future will require a more deliberate approach to systems thinking and interdisciplinary science, for example, by using frameworks that strive to characterize and integrate a broad array of interactions between humans and the environment. Although new tools and technologies can substantially improve the scientific basis of environmental policy and regulations, many of the new tools and technologies need to build on and enhance the current foundation of environmental science and engineering. Some tools and technologies that EPA has used or could use to address environmental and human health challenges are discussed in the following paragraphs.

Many advancing tools and technologies are being used to understand the transport and fate of chemicals in the environment, to understand the extent of human exposures, and to identify and predict the extent of potential toxic effects. For example, advances in separation and identification of nucleotides, proteins, and peptides and advances in spectrometric methods have enabled a better understanding of molecular-level biologic processes. Those types of tools are an integral part of EPA's computational toxicology program and are being applied to the development of new approaches to assess and predict toxicity in vitro. Advances in biomonitoring, sensor technology, health tracking, and informatics are improving the understanding of individual exposures and associated health endpoints. If EPA is to continue this work, it will need to maintain and increase its expertise in such fields as toxicology, exposure science, epidemiology, molecular biology, information technology, bioinformatics, computer science, and statistical modeling.

Advances in remote sensing since the launch of Landsat 1 in 1972 are continuing to improve the understanding of contaminant sources, fate, and transport and the understanding and monitoring of landscape ecology and ecosystem services. Using remotely collected data effectively to gain information also requires advances in modeling of various components of the Earth's biogeophysical systems, including improved techniques for data assimilation and modeling. As an example in the air-pollution arena, active sensors, such as satellite sensors and aircraft-mounted light detection and ranging sensors, can provide information on the vertical distribution of clouds and aerosols and can provide important spatial, temporal, and contextual information about the extent, duration, and transport paths of pollution. Remote sensing is also being used to monitor fugitive releases of methane, hazardous air pollutants, and volatile organic compounds from landfills and other diffuse or dispersed sources. What had been thought to be an excessively expensive monitoring challenge is proving financially and practically manageable.

Methods for identifying and quantifying chemicals, microorganisms, and microbial products in the environment continue to improve. For example, the most recent advances in the detection of microorganisms in water include quantitative polymerase chain reaction (PCR) methods, which can be designed for any microorganism of interest because they are highly specific and quantitative. In addition to updating water-quality standards and addressing health studies and swimmer surveys, EPA has begun to use PCR techniques to understand coastal pollution, address polluted sediments, decrease response time for detecting polluted waters, and improve protection of public health on beaches and coastlines. Such advances as the deployment of quantitative PCR require linking biology, mathematics, health, the environment, and policy to support substantial interdisciplinary research focused on problem-solving and systems thinking.

New tools and technologies are collecting larger, more diverse sets of data on increasing spatial and temporal scales. Knowledge and expertise in such fields as computer science, information technology, environmental modeling, and remote sensing are necessary to collect, manage, analyze, and model those datasets. One method for collecting information across larger geographic spaces and over longer periods is public engagement. For example, during massive online collaborations, participants can be invited to help to develop a new technology, carry out a design task, propose policy solutions, or capture, systematize, or analyze large amounts of data. EPA is already exploring crowdsourcing and citizen-science approaches. Improving capabilities of managing and ensuring the quality of very large datasets acquired through public engagement holds promise for EPA to be able to gather and analyze large amounts of data and input inexpensively.

Using New Science to Drive Safer Technologies and Products

The tools and technologies for handling scientific data have generally been thought of in the context of refined risk-assessment processes. That use of scientific information is focused in large part on detailed and nuanced problem identification—that is, a holistic understanding of causes and mechanisms. Such work is important and valuable in understanding how toxicants and other stressors affect environmental health and ecosystems, and at times it is required by statute. However, the focus on problem identification sometimes occurs at the expense of efforts to use scientific tools to develop safer technologies and solutions. Defining problems without a comparable effort to find solutions can diminish the value of applied research efforts. Furthermore, if EPA's actions lead to a change in a chemical, technology, or practice, there is a responsibility to understand alternatives and to support a path forward that is environmentally sound, technically feasible, and economically viable.

EPA has taken global leadership in three fields of innovative solution-oriented science: pollution prevention, Design for the Environment, and green chemistry and engineering. That suite of programs reflects non-regulatory approaches that protect the environment and human health by designing or redesigning processes and products to reduce the use and release of toxic materials. The programs emphasize education and assistance, alignment of environmental protection with economic development, and strong partnerships between agencies, industry, nongovernment organizations, and academic institutions. They require expertise in traditional environmental science, but there is also a critical need for behavioral and social sciences in advancing the development and adoption of safer chemicals, materials, and products. The data that the behavioral and social sciences provide are important inputs for characterizing and making the economic case for new technologies, for understanding business and consumer behavior, and for effecting behavioral changes so that innovations for safer materials reflect consumer preferences.

BUILDING SCIENCE AND ENGINEERING FOR ENVIRONMENTAL PROTECTION IN THE 21st CENTURY

As a regulatory agency, EPA applies many of its resources to implementing complex regulatory programs, including substantial commitments of scientific and technical resources to environmental monitoring, applied health and environmental science, risk assessment, benefit–cost analysis, and other activities that form the foundation of regulatory decisions. The primary focus on its regulatory mission can engender controversy and place strains on the conduct of EPA's scientific work in ways that do not occur in most other government science agencies. Amid this inherent tension, science in EPA generally and in EPA's Office of Research and Development (ORD) in particular strives to support the needs of the agency's present regulatory mandates and timetables, to identify and lay the intellectual foundations that will allow the agency to meet current and emerging environmental challenges, to determine the main environmental research problems on the US environmental-research landscape, to sustain and continually rejuvenate a diverse inhouse scientific staff to support the agency, and to strike an appropriate balance between inhouse and extramural research investment. In light of the inherent tensions, the current and persistent environmental challenges, and newly developed and emerging tools and technologies, the committee created a framework for building science for environmental protection in the 21st century (see Figure S-1). Environmental and human health challenges of the future and the tools and technologies that will emerge to address them cannot be predicted, but the committee offers the framework to help EPA to be prepared to respond to unknown challenges in the future and to bolster its ability to respond to current and persistent environmental challenges. The framework relies on four key ideas:

First, effective science-informed regulation and policy aimed at protecting human health and environmental quality rely on robust approaches to data acquisition, modeling, and knowledge development (see the "Analysis of Key Measures to Advance Knowledge" box in Figure S-1). Management and interpretation of "big data" will be a continuing challenge for EPA inasmuch as new technologies can generate large amounts of data quickly. In many instances, large amounts of data are acquired directly as a component of hypothesis-driven research. However, many new technologies generate large volumes of data that may not be derived from a clear, hypothesis-driven experiment but nevertheless may yield important new insights. That type of research is referred to as discovery-driven research. In both instances, the data must be analyzed and interpreted and then placed in the context of an appropriate problem or scientific theory. As depicted in Figure S-1, there must be iterations and feedback loops, particularly between data acquisition and data modeling, analysis, and synthesis. Knowledge generation, which can take many forms depending on the question being addressed and the nature of the data, ultimately serves as the basis of science-informed regulation and policy. The committee recognizes that scientific data constitute only one—albeit important—input into decision-making processes that alone cannot resolve highly complex and uncertain environmental and health problems. Ultimately, environmental health decisions and solutions will need to incorporate economic, societal, behavioral, political, and other considerations in addition to science.

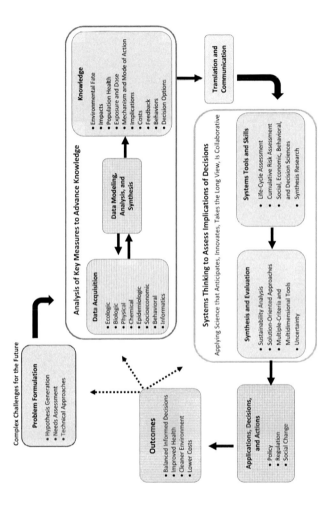

FIGURE S-1 Framework for enhanced science for environmental protection. The iterative process starts with effective problem formulation, in which policy goals and an orientation toward solutions help to determine scientific needs and the most appropriate methods. Data are acquired as needed and synthesized to generate knowledge about key outcomes. This knowledge is incorporated into an array of systems tools and solutions-oriented synthesis approaches to formulate policies that best improve public health and the environment while taking account of social and economic impacts. Once science-informed actions have been implemented, outcome evaluation can help determine whether refinements to any previous stages are required (see the dotted lines in the figure).

Second, EPA can maintain its global position by staying at the leading edge of science (see the "Systems Thinking to Assess Implications of Decisions" box in Figure S-1). Staying at the leading edge will require consideration of existing and on-the-horizon challenges and efforts to predict, address, and prevent future challenges. The committee suggests the following overarching actions for addressing wicked problems:

- *Anticipate.* Be deliberate and systematic in anticipating scientific, technologic, and regulatory challenges.
- *Innovate.* Support innovation in scientific approaches to characterize and prevent problems and to support solutions through sustainable technologies and practices.
- *Take the long view.* Track progress in ecosystem protection and human health over the medium term and the long term and identify needs for course corrections.
- *Be collaborative.* Support interdisciplinary collaboration within and outside the agency, across the United States, and globally.

Third, maintaining leading-edge science requires the development and application of systems-level tools and expertise for the systematic analysis of the health, environmental, social, and economic implications of individual decisions (see the "Systems Tools and Skills" box in Figure S-1). Leading-edge science will produce large amounts of new information, and many multifactorial problems will require systems-thinking approaches. Over the years, EPA has become more accomplished in addressing cross-media problems and avoiding "solutions" that transfer a problem from one medium to another (for example, changing an air pollutant to a water or solid-waste pollutant). However, future problems will become more complex and will go beyond cross-media situations, such as global climate and land-use patterns. Many analytic systems tools can contribute to analyzing and evaluating complex scenarios, including life-cycle assessment; cumulative risk assessment; social, economic, behavioral, and decision sciences; and synthesis research. Regardless of the analytic systems tools used, it is important to characterize and integrate information on both human health and ecosystem effects.

Fourth, maintaining leading-edge science requires the development of tools and methods for synthesizing scientific information and characterizing uncertainties. It should also integrate methods for tracking and assessing the outcomes of actions (that is, for being accountable) into the decision process from the outset (see the "Synthesis and Evaluation" box in Figure S-1). Systems-level problems are rarely amenable to simple quantitative decision measures and may require multiple types of information and characterization of different types of uncertainty. Examples of approaches for synthesizing information to support holistic decisions include sustainability analysis, solutions-oriented approaches (such as health impact assessment, alternatives assessment, and cost–benefit analysis), and multiple-criteria and multidimensional decision-making. Regardless of which analytic tools or indicators EPA uses to support decisions in the future, uncertainty will be an overriding concern. Consistent and holistic approaches to characterizing and recognizing uncertainty will allow EPA to articulate the importance of uncertainty in light of pending decisions and to avoid becoming paralyzed by the need for increasingly complex computational analysis.

The committee recommends that EPA consider the following actions to implement the elements underlying the framework in Figure S-1:

- Engage in a deliberate and systematic "scanning" capability involving staff from ORD, other program offices, and the regions. Such a dedicated and sustained "futures network" (as EPA has called groups in the past with a similar function), with time and modest resources, would be able to interact with other federal agencies, academe, and industry to identify emerging issues and bring the newest scientific approaches into EPA.
- Develop a more systematic strategy to support innovation in science, technology, and practice.
- Substantially enhance EPA's capacity to apply systems thinking to all aspects of its approach to complex decisions.
- Invest substantial effort to generate broader, deeper, and sustained support for long-term monitoring of key indicators of environmental quality and performance.

ENHANCED LEADERSHIP AND CAPACITY IN THE US ENVIRONMENTAL PROTECTION AGENCY

To implement the key strategies described above and the framework illustrated in Figure S-1, strong science leadership and capacity in EPA are essential. The committee has identified four key areas where enhanced leadership and capacity can strengthen the agency's ability to address current and emerging environmental challenges and to take advantage of new tools and technologies to address them.

Enhanced agency-wide science leadership. There has been progress toward agency-wide science integration with the establishment of the Office of the Science Advisor, and further progress might be made with the shift of the science advisor position from within ORD to the Office of the Administrator in early 2012. However, that office may need further authority from the administrator or additional staff resources to continue to improve the integration and coordination of science across the programs and regions throughout the agency. Someone in a true agency-wide science leadership position, with clear lines of authority and responsibility, could take the form of a deputy administrator for science, a chief scientist, or possibly an enhanced version of the current science advisor position. He or she could direct efforts to extend ORD's successful multiyear science plans to an *agency-wide* plan that integrates science needs of the programs and the regional offices with the scientific efforts of ORD, program offices, and regions. With such leadership in place, regional administrators, program assistant administrators, and staff members at all levels need to be held accountable for ensuring scientific quality and the integration of individual science efforts with broader efforts throughout the agency. Even with the full support of the administrator and senior staff, the effort will fail if the need to improve the use of science in EPA is not accepted by staff at all levels.

More effective coordination and integration of science efforts within the agency. Given the need for integrated, transdisciplinary, and solutions-oriented research to solve 21st century environmental problems, the existing structure focused on ORD as the "science center" that establishes the scientific agenda of EPA will not be sufficient; ORD only conducts a portion of EPA's scientific efforts, and more than three-fourths of EPA's scientific staff work outside ORD. Instead, efforts to strengthen EPA science will need to incorporate efforts, resources, expertise, and scientific and nonscientific perspectives of program and field offices. Such efforts need to support the integration of both existing and new science throughout the agency; avoid duplication or, worse, contradictory efforts; respect different sets of priorities and timeframes; and advance common goals.

Strengthened scientific capacity inside and outside the agency. Optimizing resources, creating and benefiting from scientific exchange zones, and leading innovation through transdisciplinary collaborations will require forward-thinking and resourceful scientific leadership and capacity at various levels in the agency. In such a situation, EPA would need to use all its authority effectively, including pursuing permanent Title 42 authority, to recruit, hire, and retain the high-level science and engineering leaders that it needs to maintain a strong inhouse research program. EPA would also need to maintain a "critical mass" of world-class experts who have the ability to identify and access the necessary science inside or outside EPA and to work collaboratively with researchers in other agencies. Mechanisms through which that could be achieved include sabbaticals and other leave, laboratory rotations, and the Science to Achieve Results fellowship program. The committee found that a particular area where EPA lacks expertise is in the social, behavioral, and decision sciences.

Support of scientific integrity and quality. Critics of EPA's regulations (as either too lax or too stringent) have sometimes charged that valid scientific information was ignored or suppressed, or that the scientific basis of a regulation was not adequate. EPA's best defense against such criticisms is to ensure that it distinguishes transparently between questions of science and questions of policy in its regulatory decisions; to demand openness and access to the scientific data and information on which it is relying, whether generated in or outside the agency; and to use competent, balanced, objective, and transparent procedures for selecting and weighing scientific studies, for ensuring study quality, and for peer review. The need to describe methods clearly for selecting and weighing studies is evident given the criticisms of assessments prepared for EPA's Integrated Risk Information System (IRIS). Over the last decade, several NRC committees that reviewed IRIS assessments noted a need to improve formal, evidence-based approaches to increase transparency and clarity in selecting datasets for analysis and a greater focus on uncertainty and variability. Those points were reiterated in the 2011 NRC report *Review of the Environmental Protection Agency's Draft IRIS Assessment of Formaldehyde*. EPA has announced that it is working to address the concerns raised in that report and is currently sponsoring, at the request of Congress, an NRC study to assess the scientific, technical, and process changes being implemented for IRIS.

Based on the four key areas identified above, the Committee on Science for EPA's Future recommends that EPA strengthen its capability to pursue the scientific information and tools that will be needed to meet current and future challenges by

- Substantially enhancing the responsibilities of a person in an *agency-wide* science leadership position to ensure that the highest-quality science is developed, evaluated, and applied systematically throughout the agency's programs. The person in that position should have sufficient authority and staff resources to improve the integration and coordination of science across the agency. If this enhanced leadership position is to be successful, strengthened leadership is needed throughout the agency and the improved use of science at EPA will need to be carried out by staff at all levels.
- Strengthening its scientific capacity. This can be accomplished by continuing to cultivate knowledge and expertise within the agency generally, by hiring more behavioral and decision scientists, and by drawing on scientific research and expertise from outside the agency.
- Creating a process to set priorities for improving the quality of EPA's scientific endeavors. The process should recognize the inevitably limited resources while clearly articulating the level of resources required for EPA to continue to ensure the future health and safety of humans and ecosystems.

CONCLUDING REMARKS

For over 40 years, EPA has been a national and world leader in addressing the scientific and engineering challenges of protecting the environment and human health. The agency's multi-disciplinary science workforce of 6,000 is bolstered by strong ties to academic research institutions and science advisers representing many sectors of the scientific community. A highly competitive fellowship program also provides a pipeline for future environmental science and engineering leaders and enables the agency to attract graduates who have state-of-the-art training.

The foundation of EPA science is strong, but the agency needs to continue to address numerous present and future challenges if it is to maintain its science leadership and meet its expanding mandates. There is a pressing need to groom the interdisciplinary-thinking and collaborative leaders of tomorrow and prepare for the coming retirement of large numbers of senior scientists. As this report underscores, there is an increased recognition of the need for cross-disciplinary training and of the need to expand the capacity in social and information sciences. In addition, EPA will continue to need leadership in traditional core disciplines, such as statistics, chemistry, economics, environmental engineering, ecology, toxicology, epidemiology, exposures science, and risk assessment. EPA's future success will depend on its ability to address long-standing environmental problems, its ability to recognize and respond to emerging challenges, its ability to link broader problem characterization with solutions, and its capacity to meet the scientific needs of policy-makers and the American public.

1

Introduction

The stated mission of the US Environmental Protection Agency (EPA) is to protect human health and the environment. EPA seeks to fulfill its mission by using the best available science to inform the decisions that it makes. It also seeks to ensure that federal laws related to human health and the environment are enforced fairly and effectively. The agency plays a major role in providing environmental and human health information to all members of society and works with other nations to facilitate the protection of the global environment (EPA 2011a).

EPA is carrying out its mission at a time when science is increasingly in the public eye and controversial, science budgets are decreasing, and job creation and innovation have high political priority. Science has always been an integral part of EPA's activities, and scientific assessments of factors that affect human health and the environment are as important as ever. In addition, the effects that humans continue to have on the environment are profound and widespread. An increased use of new scientific knowledge and technical information is necessary to understand increasingly complex environmental problems; to understand rapidly evolving advances in such fields as microbiology, information technology, and medicine; to set priorities for research and regulation; to identify emerging and future environmental and health concerns (NRC 2000); and to support policy, management, and technical innovations that prevent undesirable effects in the first place.

Some of the challenges and opportunities that EPA faces include new and persistent environmental problems, changes in human activities and interactions, changes in public expectations, new models for decision-making, new scientific information, and the development of new agency mission requirements that require doing more with less. EPA can meet those challenges only by using high-quality science. The present report discusses current environmental challenges and recent scientific and technologic developments, and it provides guidance to the agency as it prepares to meet the challenges of the future.

THE CHANGING NATURE OF ENVIRONMENTAL PROBLEMS

Earthquakes, floods, fires, droughts, blizzards, dust storms, natural releases of toxic gases and liquids, diseases, and other environmental variations affect hundreds of millions of people each year. Many such events are exacerbated or mitigated by human activities. In addition, humans affect the environment and natural biodiversity by adding contaminants to air and water, changing land use, reducing and fragmenting the habitat of some species, introducing non-native species, and changing natural fluxes and cycles of energy and materials. It is increasingly clear that human activities are driving many changes in Earth's global environment; indeed, some scientists refer to this human-dominated period as the Anthropocene to indicate a new geologic epoch that succeeds the Holocene. The term *Anthropocene* has also recently come into use in the popular press (for example, New York Times 2011 and The Economist 2011) and a proposal to define and formalize the term is being developed by the Anthropocene Working Group for consideration by the International Committee on Stratigraphy (SQS 2012).

The challenges associated with environmental protection today are multifaceted and affected by many interacting factors. The challenges operate on various, often large, spatial scales, unfold on long temporal scales, and usually have global implications (for example, carbon dynamics, nutrient cycles, and ocean acidification). Dealing with these problems will require systems thinking and integrated multidisciplinary science.

Achieving solutions to these challenges requires increased sustainability, the pursuit of which has been called a wicked problem. The term *wicked problem* has been used in the field of social planning to describe a problem that is difficult to solve because it is difficult to define clearly, resistant to resolution, and inadequately understood; it has multiple causes that interact in complex ways; it attracts attempted solutions that often result in unforeseen consequences; it is often not stable; it usually has no clear solution or endpoint but rather solutions that are considered better, worse, or good enough; it is socially complex and has multiple stakeholders who must consider the changing behavior of others; and it rarely sits conveniently within the understanding of one discipline or the responsibility of any one organization. Moreover, because of complex interdependencies, the effort to solve one aspect of a wicked problem may reveal or create other problems (Rittel and Webber 1973; DeGrace and Stahl 1990). There is no doubt that the environmental pollution problems of today fit the characteristics of wicked problems.

The environment is variable, complex, and difficult to predict. That difficulty is in part due to imperfect scientific knowledge about environmental processes, but it is also a consequence of imperfect knowledge about economic, demographic, and social processes that drive environmental change and the feedback effects of environmental change on economic, demographic, and social processes. Sustainable pathways to address environmental and human health challenges will only emerge if societies choose to pursue sustainable solutions and devote resources to successfully designing sustainable policies. Fully integrating sustainability as it relates to the environment and human health requires identifying and contending with tradeoffs within complex economic, cultural, and political systems. Addressing the emerging challenges that EPA faces will require not only good science and technologies, but data and information from disciplines such as social, behavioral, and decision sciences and the integration of broader frameworks that will allow a systems approach to assessing and managing issues.

Frameworks for Incorporating Human–Environment Interactions

To respond effectively to complex and rapidly changing problems, it will be important for EPA to strive toward incorporating a broader array of interactions between humans and the environment into its regulatory and decision-making processes, identify optimal ways to advance core human development and sustainability goals, understand the tradeoffs that necessarily accompany decisions about specific ways to use environmental resources, and align response options with the level of governance at which options can be most effective. Several frameworks have been developed to identify and incorporate the full array of interactions between humans and the natural environment into planning and evaluation. The framework proposed by the Millennium Ecosystem Assessment (MEA) (MEA 2003, 2005) is useful because it includes the intrinsic value of biodiversity and ecosystems and recognizes that people use multiple criteria when making decisions about how to use the environment. The MEA framework focuses particular attention on the linkages between ecosystem services and human well-being (Figure 1-1) and also stresses the roles of science and engineering as direct and indirect drivers of environmental change. Similar frameworks have been developed by committees of the National Research Council (NRC) (NRC 2000, 2004) and EPA's Science Advisory Board (EPA SAB 2002, 2009). The Heinz Center (2002, 2008) also developed a comprehensive framework for assessing the state of the nation's ecosystems.

The frameworks highlight the importance of a comprehensive conceptual model of the environmental system that includes its structural elements, compositional elements, and dynamic functional properties. They also all direct attention to the supporting services (primary production, nutrient cycling, and soil formation) that are necessary for the generation of all other ecosystem services. EPA can draw upon those frameworks and increase its use of systems thinking as it incorporates new knowledge and technical tools into its science and management activities. Taking advantage of those types of frameworks will require scientific consortia that can provide an improved understanding of the problem, create opportunities for interactions between diverse areas of specialization, and integrate knowledge to identify effective solutions. This is a large job for any single agency or organization, so it will be imperative that networks and partnerships be created or enhanced. It will also be necessary for EPA to communicate with a wide range of experts, particularly for integrating emerging work in social sciences and information technology with advances in exposure assessment and risk assessment.

SCIENCE AND ENGINEERING AT THE US ENVIRONMENTAL PROTECTION AGENCY

EPA has been aware of the implications of the rapid growth of scientific data, concepts, and technical tools and has begun to incorporate many scientific advances into its major activities. It has also made substantial efforts to comprehend the unprecedented complexities of emerging environmental problems and to prepare to respond appropriately to the challenges that these developments pose for both its research and its regulatory responsibilities. However, because EPA is a regulatory agency and is not fundamentally a science agency, the role EPA plays supporting science to protect the environment and human health can sometimes be challenging.

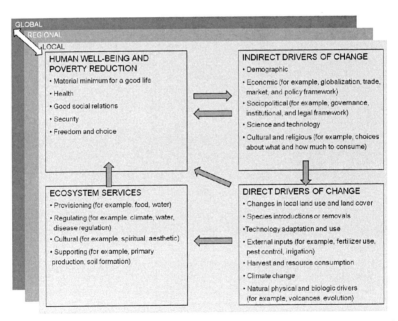

FIGURE 1-1 The Millennium Ecosystem Assessment conceptual framework. Indirect drivers of change (such as demographics, economic factors, science, and technology) can cause changes in ecosystems, which in turn can have direct effects on human well-being. These interactions can exist on local, regional, and global scales and can cause changes in both the short term and long term. Direct and indirect feedbacks among drivers are common. For more information on this particular framework, see MEA 2003 and MEA 2005. Source: Adapted from MEA 2003.

Since its formation in 1970, EPA has played a leadership role in developing many fields of environmental science and engineering, from ecology to health sciences and environmental engineering to analytic chemistry. EPA has performed, supported, and stimulated academic research; developed environmental education programs; supported regional science initiatives; supported the development and application of new technologies; and, most important, enhanced the scientific information that creates a basis for regulatory decisions (NRC 2000, 2003; Collins et al. 2008; Darnall et al. 2008; Kyle et al. 2008; Sanchez et al. 2008; NRC 2011). The broad reach of EPA science has also influenced international policies and guided state and local actions. Some examples of traditional EPA science-based and engineering-based initiatives are identifying emerging ecologic and health problems, monitoring trends in ecologic systems and pollution, identifying human health hazards, measuring and modeling population exposures, developing pollution-control technologies, supporting health-based enforcement and standard-setting, tracking environmental improvement, and incorporating green chemistry concepts and pollution prevention solutions.

Environmental Protection Agency Successes

EPA has successfully contributed to the reduction of pollution and improved public health, human welfare, and environmental and ecosystem quality. Its success has stemmed largely from the establishment and enforcement of its regulatory programs under the Safe Drinking Water Act; the Clean Water Act; the Clean Air Act; the Federal Insecticide, Fungicide, and Rodenticide Act; the Comprehensive Environmental Response, Compensation, and Liability Act (also known as Superfund); the Toxic Substances Control Act; and other statutes. Such success would not be possible without scientific and engineering support within the agency and outside by universities, colleges, and partnering agencies and companies. An example of EPA's success involves the regulation of air pollutants. Many conventional air pollutants have been dramatically reduced over a 20-year period (Figure 1-2)—a demonstration of the remarkable success that the United States has achieved by amending and enforcing the Clean Air Act. It is expensive to implement the Clean Air Act, but it has resulted in improved economic welfare, including better health, improved labor productivity, and less morbidity and mortality due to air pollution (EPA 2011b).

As shown in Table 1-1, there have been large declines in the emissions of nitrogen oxide gases, volatile organic compounds, carbon monoxide, sulfur dioxide, lead, and particulate matter smaller than 10 µm in diameter and smaller than 2.5 µm in diameter over the last 30 years. Despite a doubling of the US gross domestic product during that period and large increases in vehicle-miles traveled, population, energy consumption, and carbon dioxide emissions, regulation of the transportation and industrial sectors has reduced emissions of conventional air pollutants and brought about cleaner air (see Figure 1-2).

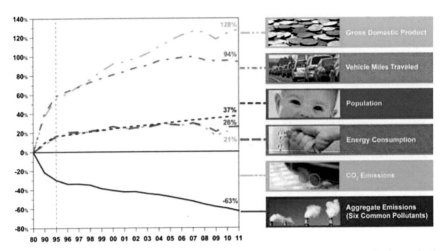

FIGURE 1-2 Gross trends in drivers and aggregate emissions since 1980 in the United States. Source: EPA 2012a.

TABLE 1-1 Change in Conventional Air Pollutant Emissions Over the Last 3 Decades

	Change, %		
	1980–2010	1990–2010	2000–2010
Carbon monoxide (CO)	-71	-60	-44
Lead (Pb)	-97	-60	-33
Nitrogen oxides (NO_x)	-52	-48	-41
Volatile organic compounds (VOCs)	-63	-52	-35
Direct particulate matter less than 10 μm in diameter (PM_{10})	-83[a]	-67	-50
Direct particulate matter less than 2.5 μm in diameter ($PM_{2.5}$)	---[b]	-55	-55
Sulfur dioxide (SO_2)	-69	-65	-50

[a] Direct PM_{10} emissions for 1980 are based on data since 1985.
[b] --- Trend data not available.
Source: EPA 2012a.

Introduction   *21*

In 2010, in recognition of the agency's 40th anniversary, a distinguished group of environmental professionals representing government, nongovernment organizations, and the private sector assembled to identify EPA's key achievements (Aspen Institute 2010). The list included removing lead from gasoline to improve air quality and children's health, reducing acid rain to improve water quality in lakes and streams, reducing exposure to second-hand smoke by identifying environmental tobacco smoke as a human carcinogen, spurring improvements in vehicle efficiency and emission control, testing requirements and encouraging "green chemistry", banning widespread use of dichlorodiphenyltrichloroethane (DDT), encouraging a shift to rethinking of waste as materials, and highlighting concerns about environmental justice. EPA scientists and engineers have been at the center of each of those accomplishments, developing cutting-edge tools for modeling and monitoring natural and engineered environmental systems, designing regulatory approaches to encourage private-sector innovation, and interpreting health and ecosystem science that is generated by external sources to inform policy decisions (EPA 2012b,c).

EPA's role in advancing environmental science and engineering continues. The agency leads research and development efforts, such as codevelopment of a system that provides early warning for water utilities to detect potential contamination (EPA 2011c). The agency is leading efforts to transform chemical toxicity testing by developing a cutting-edge computational toxicology center via unprecedented trans-federal collaborations with the National Institutes of Health (especially the National Institute of Environmental Health Sciences and the National Toxicology Program) and the Food and Drug Administration (EPA 2012d). This interagency cooperation has resulted in the development of Tox21. The agency also leads work with Canada to assess the condition and protection of the Great Lakes (EPA 2009). EPA is the only major agency that is supporting the development of new molecular methods for assessing viruses in groundwater, *Cryptosporidium* and other emerging pathogens in water, and microbial source tracking tools for addressing impairment. And EPA continues to play a leading role internationally in advancing the scientific understanding of continental-scale and global-scale atmospheric chemistry and transport with recent efforts to refine models for short-term forecast applications and efforts to understand how air-quality problems might be affected by long-term climate change.

Challenges Facing the Environmental Protection Agency

EPA scientists and engineers are addressing some of the nation's most complex technical challenges, such as standard-setting for chemical pollutants, dealing with emerging waterborne pathogens, and protection of air and water resources. Owing to its legislative mandates, EPA investigations are often initiated in response to a crisis or new information that identifies a hazard to human health or the environment. Much of EPA's science has been reactive, addressing problems after they have become widespread and focusing on cleanup or "end of pipe" solutions, rather than proactive and oriented toward long-term goals that will help the agency to address and possibly prevent environmental problems in the future.

Today, despite its considerable successes, science at EPA is facing unprecedented challenges. An NRC report, *Science and Decisions: Advancing Risk Assessment*, identified new approaches to formulate environmental problems, assess risks, and evaluate decision options (NRC 2009), which would facilitate systems thinking and innovative problem-solving discussed in the current report. Another recent NRC report, *Sustainability and the U.S. EPA*, identified broader tools incorporating economics and social sciences for evaluating decision options and formulating research programs (NRC 2011). By acknowledging past achievements and current efforts but also recognizing the many challenges that EPA faces, the current report seeks to provide advice on the initiation of new directions and approaches for science at EPA to ensure that the agency continues to generate and make effective use of the world-class science and engineering that are needed to accomplish its mission. Specific challenges that EPA faces today and will likely face in the future and tools and technologies to address them are elaborated on in Chapters 2 and 3 of this report.

THE COMMITTEE'S TASK

EPA asked NRC to assess independently the overall capabilities of the agency to develop, obtain, and use the best available scientific and technologic information and tools to meet persistent, emerging, and future mission challenges and opportunities. Those challenges and opportunities include new and persistent environmental problems, changes in human activities and interactions, changes in public expectations, new risk-assessment and risk-management paradigms, new models for decision-making, and new agency mission requirements. EPA asked that special consideration be given to a potentially increasing emphasis on transdisciplinary approaches, systems-based problem-solving, scientific and technologic innovation, and greater involvement of communities and stakeholders. NRC was also asked to identify and assess transitional options to strengthen the agency's ability to pursue the aforementioned scientific information and tools. In response, it convened the Committee on Science for EPA's Future, which prepared the present report. The committee's full statement of task is provided in Appendix A, and biographic information on the committee is in Appendix B.

To accomplish its task, the committee held six meetings from June 2011 to April 2012. The first two meetings included public sessions during which the committee heard from several EPA staff and from a principal investigator at the National Institute of Environmental Health Sciences. In writing its report, the committee gathered information through communication with EPA staff, from resources on EPA's website, peer-reviewed scientific literature, and reviews and reports written by numerous other government agencies, nongovernment organizations, and independent advisory groups.

ORGANIZATION OF THE REPORT

The committee's report covers a broad array of topics that reflect EPA's expansive scope to protect human health and the environment and its leadership role in local, state, and international science. In addition to EPA's need to provide scientific information that will act as the basis of regulatory decision-making, it plays a role in stimulating and supporting academic research, environmental-education programs, and regional science initiatives and in providing support for safer technologies. Science is needed to support EPA as both a regulatory agency and as a leader in environmental science and engineering. While this report focuses on the issues of science, data, and information management, it recognizes that the policy changes facing EPA and environmental protection more broadly are important.

This report is organized into six chapters and four appendixes. Chapter 2 discusses persistent challenges that EPA is facing now and emerging challenges that may be important to EPA in the future. In the context of those challenges, Chapter 3 aims to provide information on emerging tools and technologies for environmental protection and the application of those emerging tools and technologies. Chapter 4 addresses approaches for EPA to remain at the leading edge of environmental science and engineering, to evaluate and synthesize leading-edge science to inform decisions, to deliver science within and outside the agency, and to strengthen its science capacity. Specific details related to "-omics" technologies and information technology are elaborated on in Appendixes C and D, respectively. Chapter 5 specifically addresses enhanced science leadership and scientific capacity at EPA. Chapter 6 summarizes the committee's main findings and recommendations.

The committee uses the word *science* in this report in two distinctive ways. One refers to the processes—collectively called the scientific method—by which new information is generated (that is, research). The second way refers to the body of knowledge produced by scientific methods—that is, the resulting data. EPA both conducts high-quality research and uses scientifically generated information in many ways. The challenges and tools and technologies that the committee discusses are meant to be examples of the types of problems EPA faces now, the types of problems EPA could potentially face in the future, and the types of tools and technologies that could help to solve current, persistent, and emerging environmental challenges. The committee cannot anticipate all of the problems of the future and the tools and technologies that will be needed to address those problems, so it has focused on describing a framework that will help EPA to be better prepared in the future. Some of the committee's findings and recommendations concern the agency's science programs, and many are related to EPA's role in synthesizing data to inform policy decisions and the establishment of regulations, and to stimulate thinking in new ways. The mechanism or mechanisms through which EPA chooses to address the recommendations will depend on its funding, its priorities, and what environmental science and engineering areas it wants to focus its efforts on in the future. Because the committee's report will become dated as science evolves and as lessons continue to be learned about best practices for protecting human health and the environment, it may be beneficial for EPA to carry out a similar type of exercise at regular intervals in the future.

REFERENCES

Aspen Institute. 2010. EPA 40th Anniversary: 10 Ways EPA has Strengthened America. Aspen Institute, November 2010 [online]. Available: http://www.aspeninstitute.org/sites/default/files/content/docs/events/EPA_40_Brochure.pdf [accessed Nov. 18, 2011].

Collins, F. S., G.M. Gray, and J.R. Bucher. 2008. Transforming environmental health protection. Science 319(5865):906-907.

Darnall, N., G.J. Jolley, and R. Handfield. 2008. Environmental management systems and green supply chain management: Complements for sustainability? Bus. Strat. Env. 18(1):30-45.

DeGrace, P., and L.H. Stahl. 1990. Wicked Problems, Righteous Solutions: A Catalog of Modern Engineering Paradigms, 1st Ed. Englewood Cliffs, NJ: Yourdon Press.

EPA (US Environmental Protection Agency). 2009. The Great Lakes Water Quality Agreement. US Environmental Protection Agency [online]. Available: http://www.epa.gov/greatlakes/glwqa/usreport/part6.html [accessed Jan. 20, 2012].

EPA (US Environmental Protection Agency). 2011a. Our Mission and What We Do. US Environmental Protection Agency [online]. Available: http://www.epa.gov/aboutepa/whatwedo.html [accessed Mar. 19, 2012].

EPA (US Environmental Protection Agency). 2011b. The Benefits and Costs of the Clean Air Act from 1990 to 2020. Office of Air and Radiation, U.S. Environmental Protection Agency, Washington, DC. March 2011 [online]. Available: http://www.epa.gov/oar/sect812/feb11/fullreport.pdf [accessed July 21, 2012].

EPA (US Environmental Protection Agency). 2011c. Enhancing Water Security: EPA Prepares for Intentional Contamination Incidents. Science Matters Newsletter. US Environmental Protection Agency [online]. Available: http://www.epa.gov/sciencematters/september2011/contamination.htm [accessed Jan. 20, 2012].

EPA (U.S. Environmental Protection Agency). 2012a. Air Quality Trends [online]. Available: http://www.epa.gov/airtrends/aqtrends.html#comparison [accessed July 30, 2012].

EPA (US Environmental Protection Agency). 2012b. Economic Incentives. National Center for Environmental Economics, US Environmental Protection Agency [online]. Available: http://yosemite.epa.gov/ee/epa/eed.nsf/pages/EconomicIncentives.html [accessed Jan. 20, 2012].

EPA (US Environmental Protection Agency). 2012c. SPARC. US Environmental Protection Agency [online]. Available: http://www.epa.gov/athens/research/projects/sparc/ [accessed Mar. 27, 2012].

EPA (US Environmental Protection Agency). 2012d. Computational Toxicology Research Program. US Environmental Protection Agency [online]. Available: http://www.epa.gov/ncct/ [accessed Feb. 4, 2012].

EPA SAB (US Environmental Protection Agency Science Advisory Board). 2002. A Framework for Assessing and Reporting on Ecological Condition: Executive Summary. EPA-SAB-EPEC-02-009A. US Environmental Protection Agency Science Advisory Board, Washington, DC. September 2002 [online]. Available: http://yosemite.epa.gov/sab/sabproduct.nsf/CB5152C32D7EE6278525710000683FD3/$File/epec02009a.pdf [accessed Mar. 27, 2012].

EPA SAB (US Environmental Protection Agency Science Advisory Board). 2009. Valuing the Protection of Evological Systems and Services: A Report of the EPA Science Advisory Board. EPA-SAB-09-012. US Environmental Protection Agency Science Advisory Board, Washington, DC. May 2009 [online]. Available: http://yosemite.epa.gov/sab/sabproduct.nsf/WebBOARD/SAB-09-012/$File/SAB%20Advisory%20Report%20full%20web.pdf [accessed July 21, 2012].

Heinz Center (The H. John Heinz III Center for Science, Economics and the Environment). 2002. The State of the Nation's Ecosystems: Measuring the Lands, Waters, and Living Resources of the United States. Cambridge, UK: Cambridge University Press.

Heinz Center (The H. John Heinz III Center for Science, Economics and the Environment). 2008. The State of the Nation's Ecosystems 2008: Measuring the Lands, Waters, and Living Resources of the United States. Washington, DC: Island Press.

Kyle, J.W., J.K. Hammitt, H.W. Lim, A.C. Geller, L.H. Hall-Jordan, E.W. Maibach, E.C. De Fabo, and M.C. Wagner. 2008. Economic evaluation of the U.S. Environmental Protection Agency's SunWise Program: Sun protection education for young children. Pediatrics 121(5): e1074-e1084.

MEA (Millennium Ecosystem Assessment). 2003. Ecosystems and Human Well-Being: A Framework for Assessment. Washington, DC: Island Press [online]. Available: http://pdf.wri.org/ecosystems_human_wellbeing.pdf [accessed July 30, 2012].

MEA (Millennium Ecosystem Assessment). 2005. Ecosystems and Human Well-Being: Synthesis. Washington, DC: Island Press [online]. Available: http://www.maweb.org/documents/document.356.aspx.pdf [accessed July 30, 2012].

New York Times. 2011. Editorial: The Anthropocene. New York Times, February 27, 2011[online]. Available: http://www.nytimes.com/2011/02/28/opinion/28mon4.html?_r=2 [accessed Nov. 18, 2011].

NRC (National Research Council). 2000. Ecological Indicators for the Nation. Washington, DC: National Academy Press.

NRC (National Research Council). 2003. The Measure of STAR: Review of the US Environmental Protection Agency's Science to Achieve Results (STAR) Research Grants Program. Washington, DC: National Academies Press.

NRC (National Research Council). 2004. Valuing Ecosystem Services: Toward Better Environmental Decision-Making. Washington, DC: National Academies Press.

NRC (National Research Council). 2009. Science and Decisions: Advancing Risk Assessment. Washington, DC: National Academies Press.

NRC (National Research Council). 2011. Sustainability and the US EPA. Washington, DC: National Academies Press.

Rittel, H., and M. Webber. 1973. Dilemmas in a general theory of planning. Policy Sci. 4(2):155-169.

Sanchez, M.C., R.E. Brown, C. Webber, and G.K. Homan. 2008. Savings estimates for the United States Environmental Protection Agency's ENERGY STAR voluntary product labeling program. Energ. Policy 36(6):2098-2108.

SQS (Subcommission on Quaternary Stratigraphy). 2012. Working Group on the 'Anthropopocene'. Subcommission on Quaternary Stratigraphy.[online]. Available:

http://www.quaternary.stratigraphy.org.uk/workinggroups/anthropocene/ [accessed July 22, 2012].

The Economist. 2011. The Anthropocene: A Man-Made World. The Economist, May 26, 2011 [online]. Available: http://www.economist.com/node/18741749 [accessed Nov. 17, 2011].

2

Challenges of the 21st Century

Efforts of the US Environmental Protection Agency (EPA) to address environmental degradation over the last 40 years have had some marked successes, including reductions in particulate and sulfur air pollution, reductions in industrial discharges in waterways, and removal of lead from gasoline. Yet enormous challenges remain. Although many of the more visible environmental problems have been at least partly addressed, persistent problems and new problems affect the environment's ability to provide the ecosystem services on which humans and other living organisms depend.

Solving current environmental challenges—for example, nutrient overload and eutrophication, climate change, increased body burdens of diverse chemicals, and water-quality declines—requires understanding the nature of the problems and their relationships to other phenomena. In particular, solving environmental challenges requires consideration of root causes and possible unintended consequences of interventions in domains not normally considered. Developing a strong understanding of how various key drivers can affect multiple phenomena relies on the expansive application of systems thinking. Identifying viable and sustainable solutions that will optimize economic, social, and environmental benefits should have high priority. Ensuring that EPA has the scientific capacity to promote those solutions requires a science strategy that builds on accomplishments but includes innovative and diverse tools.

Current and future environmental challenges also include disasters, which require EPA to have an ability to respond quickly to address environmental consequences. Those disasters can arise from natural events such as storms, earthquakes, and volcanic eruptions; from accidents at major industrial facilities, such as pipelines, large bulk-storage facilities, mines and wells, and power and chemical plants; or as the direct or indirect consequence of terrorism events. EPA is and will continue to be responsible for monitoring and addressing the environmental changes resulting from disasters (whether natural or human-caused).

Chapter 2 discusses major factors that lead to environmental change and some of the persistent challenges that EPA will likely continue to face in the coming decades. The committee cannot predict with certainty what new environmental problems EPA will face in the next 10 years or more, but it can identify some of the common drivers and common characteristics of problems. The specific topics discussed in this chapter were identified based on committee expertise and a review of the scientific literature. This chapter is not meant to be an exhaustive list of all factors leading to environmental changes or of all persistent and future environmental challenges. Instead, the chapter is meant to provide some illustrative examples of the types of problems facing EPA today and some of the factors that create and influence those problems.

MAJOR FACTORS LEADING TO ENVIRONMENTAL CHANGE

Major socioeconomic factors are directly and indirectly driving environmental changes and are increasing the imperative for EPA to maintain and strengthen its environmental research efforts. Those socioeconomic factors are often reflected in population growth and migration, demographic shifts, land-use change and habitat loss, increasing energy demand and shifting energy supplies, new consumer technologies and consumption patterns, increasing emissions of greenhouse gases, and movement of organisms beyond their traditional ranges, which in turn have implications for the scientific knowledge that is required to inform policy decisions at EPA effectively. EPA will be challenged in coming years to adapt to rapid changes in scientific knowledge, society, and the environment. An increased awareness of the effects of human activity on human health and the environment has raised people's concern regarding the issues that the agency is charged with addressing.

Population Growth

It took until 1800 AD for mankind to reach a population of 1 billion people, but only required 123 more years to reach 2 billion, 33 more years to reach 3 billion, and about 13–14 more years for each additional billion people thereafter (UN 1999). In October 2011, the worldwide population hit 7 billion (UN 2011). With the dramatic increase in population, human activities have altered and will continue to alter an ever-increasing portion of Earth's surface (Wulder et al. 2012). Such activities have diminished natural ecosystems and the benefits that they provide, including water purification, flood control, climate moderation, and new crop plants.

In the United States, the population continues to increase at approximately 1% per year (US Census Bureau 2012). This population growth contributes to such environmental effects as increased emissions of greenhouse gases due to energy use, transportation demand, and residential and commercial activities (EPA 2011a); increased consumption of resources (Worldwatch Institute 2011);

increasing numbers of manufactured chemicals and products introduced into the environment (EPA 2011a); and increased food and water demand and concomitant changes in land use (NRC 2011). Those demographic, consumption, and production changes contribute to the challenge of addressing environmental problems and health outcomes as increasing amounts of land and resources are demanded to meet human wants and needs.

Changes in Land Use

Land use is a major factor driving environmental quality. Land use strongly influences water quality through runoff, water quantity through influence on the hydrologic cycle, air quality through emissions and deposition and carbon storage in terrestrial landscapes, and biologic diversity through habitat loss, disturbance, and resource availability. In the United States, changes in land use result largely from expansion of urban and agricultural areas, energy development, and changes in forestry practices.

Population growth and demographic transitions have increased the requirement of land area for residential, commercial, and transportation activities (Squires 2002). In the conterminous United States, it has been estimated that up to 45.5 million acres (2.4%) of land is characterized by impervious surfaces (including roads, building, sidewalks, and parking lots) (Nowak and Greenfield 2012). Impervious surfaces change the hydrology and ecology of rivers (higher peak flows and scouring of habitat) and reduce the availability of groundwater for agriculture and other human use. In addition, the interconnected effects of urban sprawl are numerous and complex—greater automobile use in less-densely populated communities can lead to increased air pollution and more sedentary lifestyles, both of which are risk factors for heart disease. Less dense housing also increases energy use per capita and contributes to increased air pollution and climate change and potentially to such adverse health effects as increased asthmatic attacks (Frumkin 2002; Younger et al. 2008; Brownstone and Golob 2009).

Despite increased demand for food and fuel, the land area dedicated to agriculture has not increased substantially over the last few decades. In the United States, acreage devoted to corn has increased over the last 10 years, but total agricultural acreage has been largely unchanged. Agricultural productivity has increased as a result of major investments in research by both the public and private sectors, but there is still uncertainty as to whether the increase can be maintained and, if so, whether it would have associated environmental costs. For example, without substantially increased nutrient-use efficiency, increased amounts of fertilizers will be applied per acre of agricultural land, and therefore increased amounts of those nutrients will be lost to the environment. If increased productivity is not maintained, more acres will need to be devoted to agriculture, probably at the expense of marginally productive lands and natural ecosystems.

Increased demand for bioenergy, wind, and solar-power plants may also place additional pressure on land resources. Beyond ethanol-based biofuels, much of the bioenergy used in power generation is likely to come from forest biomass through increased use of harvesting residues and (potentially) increased harvesting. Forest ownership patterns have shifted over the last 20 years as a result of the large-scale disaggregation of the forest-products industry. That shift has increased land-use decisions that are based on maximizing shorter-term economic returns rather than long-term production of forest products (USDA 2006). When combined with more intensive use of forests to meet the demand for a shifting basket of products (largely bioenergy), shifts in forest ownership may have increasing effects on the environment. Thus, to pursue its environmental-protection mission effectively in coming years, EPA will need to expand its efforts to monitor and understand land-use changes.

Energy Choices

Energy choices in the United States—including bioenergy, conventional and unconventional oil and gas production, coal, and nuclear power—all have important implications for the environment through the effects of resource extraction or production, fuel combustion, and waste discharge or disposal. The April 2010 blowout of British Petroleum's Macondo deepwater oil well illustrated how devastating the unintended consequences of energy development can be; the accident killed 11 workers and led to the largest oil spill in US history and the closure of some fisheries in more than 80,000 square miles of the Gulf of Mexico (NOAA 2012a). The rapid but less dramatic expansion of natural-gas production across the United States has raised concerns about effects on local water and air quality. There are also concerns about greenhouse-gas emissions associated with methane leakage during production and transport, although natural gas is recognized as a fuel that inherently emits less greenhouse gas (about half) than coal when combusted (Jaramillo et al. 2007). The comparative advantages are lost at higher leak rates (that is, the rate at which methane, the primary constituent of natural gas, is lost to the atmosphere during the production, transportation, and use of natural gas) (Alvarez et al. 2012).

Another example is the production of ethanol for use as a biofuel, which has increased rapidly in the last decade because of the desire for energy security and renewable transportation fuels. In 2010, about 40% of US corn production was used as feedstock for biofuel production (NRC 2011). Such agricultural and energy choice practices can have negative environmental effects; increased production of corn as an ethanol feedstock has resulted in increased nutrient runoff and corresponding eutrophication of coastal waters, including the Gulf of Mexico (NRC 2008, 2011). Given current water-use efficiencies, large quantities of water are also required for irrigation and the intensification of agricultural practices can increase erosion (NRC 2008, 2011). Further research is required to develop new perennial feedstocks that would require less tillage and have high nutrient-use efficiencies so that soils and nutrients would be held in place. Ultimately, competition between the demand for food and the demand for land needed for other purposes will limit the amount of biofuels that can be produced. The extent to which new technology can alleviate those constraints is unclear because of limitations in photosynthetic efficiency. An improved understanding at EPA of the potential effects of new energy options and emerging technologies would help ensure that they are pursued in ways that protect the environment and human health. Broadly, the domain of energy is a classic example where systems thinking would be needed, as technologic or regulatory changes influencing one fuel type can have ripple effects across the life cycle of multiple fuels. For example, emissions requirements on power plants could reduce air pollutant emissions from coal-fired power plants and decrease impacts related to coal mining and transport, but could lead to increased use of natural gas and hydrofracturing as an extraction technology. Systems-level analyses that take account of these ripple effects and determine the net implications for ecologic and human populations are crucial.

Technologic Change and Changing Consumption

Technologic innovation creates a large challenge to acquiring the environmental data required to inform policy in a timely way. In the last 2 decades, a revolution in electronics has led to such devices as cellular telephones, iPods, and tablet computers. In 1980, the computer-chip industry used only 11 elements from the rare earth and platinum series metals; today it requires 60 elements, or almost two-thirds of the natural periodic table (Schmitz and Graedel 2010; Erdmann and Graedel 2011). Such technologic change not only requires increasing production but challenges the ability of industry to recycle and recover the (sometimes toxic) materials used in electronic devices. EPA is challenged to assimilate or perform research fast enough to understand the health and environmental risks associated with the production and disposal of those devices, let alone how to mitigate any risks. A legacy of contaminated soils in both terrestrial and aquatic environments is a reminder that managing these technologic challenges is not new. Increased vigilance is necessary to ensure that future generations are not left with a legacy of contamination as has happened in the past.

Other innovative technologies—such as new chemicals, nanomaterials, and synthetic biology—are important for economic growth. However, they also require focused research to understand adverse human health and environmental effects and to understand how to avoid harmful effects through safe product design and to ensure that wastes are reused or recycled. In the face of rapid technologic innovation, a key challenge for EPA is acquiring the scientific data required to fulfill its mission of protecting human health and the environment without imposing a drag on economic development (see Chapter 4). Understanding how new technologies will influence the application and use of existing technologies will be important in ensuring the net benefit of EPA's efforts. Social-science and behavioral-science research will be critical in helping to design and evaluate strategies for meeting that challenge.

Transport of Organisms

The geologically recent evolution of life occurred on isolated continents, each of which evolved a distinctive biota. However, the ever-expanding movement of people and goods has tended to homogenize Earth's biota and resulted in two increasingly serious environmental problems: the spread of animal-vectored diseases and the invasion of exotic species. Species are transported around the world inadvertently on ships, airplanes, and automobiles. Others are deliberately imported for agriculture, horticulture, biologic control, and recreation (such as pets or game animals). Most do not become established in the locations to which they are introduced, and few of the ones that do naturalize disrupt the local ecologic communities seriously. However, some do become highly invasive, dominating ecologic communities, spreading diseases, and diminishing the ability of other species to survive. One example is the impact zebra mussels have had in the Great Lakes region (Pejchar and Mooney 2009). Zebra mussels compete with some fish for zooplankton prey, clog intake pipes and impair flow at water treatment plants, contribute to the bioaccumulation of mercury and lead, and change nutrient balances in the water resulting in increased phytoplankton and cyanobacterial blooms. Few studies have been done to try to estimate the total costs of nonnative invading species at a national level; however, one study estimates that about $120 billion is spent in the United States per year due to environmental damages and losses caused by nonnative invading species (Pimentel et al. 2005; Pejchar and Mooney 2009). Increasingly, people are introduced to new exposure pathways and vectors through other animals that are potential carriers of diseases to which humans and other animals lack immunity.

ENVIRONMENTAL AND HUMAN HEALTH CHALLENGES

The patterns of change briefly described above have resulted in a suite of current and emerging environmental and human health challenges for EPA, such as

- Human and environmental exposure to increasing numbers, concentrations, and types of chemicals. Factors contributing to human and environmental exposures include energy choices, technologic change, and changing energy consumption.
- Threat of deteriorating air quality through changes in weather (Jacob and Winner 2009) and through the formation of more particles in the atmosphere from allergens, mold spores, pollen, and reactions of primary air pollutants

(Confalonieri et al. 2007). Factors contributing to deteriorating air quality include population growth, energy choices, changing consumption, and climate change.

- Water quality and coastal-system degradation, including challenges to rebuild old infrastructure and address such issues as urban stormwater and bypass of raw sewage (NOAA 2012b). Factors contributing to water quality and coastal-system degradation include land use, urban sprawl, climate change, and energy systems.
- Non–point-source pollution and nutrient effects associated with agricultural runoff of nutrients and soils. Factors contributing to non–point-source pollution and nutrient effects include climate change, land use, and technologic change (NRC 2011).
- Expanding quantities of waste with a wider array of component materials (Schmitz and Graedel 2010). Factors contributing to expanding quantities of waste include population growth, energy usage, technologic change, and changing consumption.
- Expanding ecologic disruptions (USDA 2012). Factors contributing to ecologic disruptions include population growth, land use, climate change, and transport of organisms.

The first three of the challenges listed above are discussed in greater detail below, with some examples that illustrate the need for a better approach for accessing, obtaining, developing, and using science and engineering in the pursuit of environmental solutions. In addition, an overarching challenge relates to the ever expanding spatial and temporal scales at which many of these challenges operate. Although the challenges in this chapter are only illustrative of today's challenges and although it is difficult to predict what emerging challenges will dominate in the future and what global implications will arise from local-scale environmental drivers, it is quite likely that future emerging challenges will share key features of the examples below. Some of those key features include complex feedback loops, the need to understand the effects of low-level exposures to numerous stressors rather than high-level exposures to individual stressors, and the need for systems thinking to devise optimal solutions.

Chemical Exposures, Human Health, and the Environment

Human health is inextricably linked with ecosystems and the quality of the environment. Since the beginnings of the discipline of public health, it has been recognized that most diseases are influenced by three factors: the agent (chemical, biologic, or physical), the host (genetic or behavioral), and the environment (physical or social). Historically, the greatest advances in controlling infectious diseases have been based on environmental improvements, such as improvements in water quality, sewage treatment, and food protection. Controlling chemical exposures and reducing or preventing associated health effects can be more challenging.

Although new chemicals continue to be created and enter the environment, many of the problems they cause are not new. Cancer was among the dominant health concerns through the early decades of EPA. Carcinogenic pollutants—including asbestos, arsenic, benzene, hexavalent chromium, dioxin, and vinyl chloride—were a major focus of interest in human health effects because of both public concerns and expanded toxicologic and epidemiologic findings. Identifying and controlling carcinogens was a dominant driver of EPA science, from analytic chemistry through toxicity testing and risk assessment. While cancer will continue to be an EPA and societal priority, other health outcomes are likely to receive increasing attention given growing epidemiologic and toxicologic evidence. Many of these health effects are chronic and subtle, and there is still much to be learned. For example, hormonally active chemicals have long been researched, but the importance of their potential health effects continues to be elucidated. A new class of hormonally active substances receiving increased attention are obesogens, which target lipid metabolism and may interfere with natural hormone signaling (Kirchner et al. 2010).

Another challenge related to exposure to chemicals or other stressors is characterizing susceptibility to adverse health effects. Susceptibility can vary greatly in a population as a function of factors that are not often systematically evaluated. Young children may be at greater risk for neurologic and endocrine effects, and the elderly may be more susceptible to immune effects, cardiovascular effects, or infection. Race or socioeconomic status may increase the risk of cumulative environmental effects that result from living disproportionally closer to pollution sources (Bullard 2000). Poverty, stress, and lack of access to medical care decrease human resilience and the ability to adapt; disadvantaged communities are at increased risk when faced with increased exposure. Genetic factors also influence susceptibility and underscore the importance of gene-environment interaction in determining health outcomes.

Transgenerational effects and sensitive populations are also of great concern for public health. Exposure to chemicals and other stressors during gestation can affect the mother, the fetus, and even the germ cells of the fetus and lead to effects on the third generation (Holloway et al. 2007). Some research indicates that chemical exposure in the womb can trigger epigenetic changes much later in life. Adipose-tissue development, food intake, and lipid metabolism may be altered as a result of exposure to organotins, perfluorooctanoic acid, diisobutyl phthalate, bisphenol A, and other xenobiotic chemicals found in the environment (Grun and Blumberg 2006). The epidemic of obesity, diabetes, and metabolic syndrome in the United States and elsewhere indicates that research is needed to determine whether there is a causal link to the chemicals described above at concentrations measured in the environment. If environmental exposures caused even a tiny fraction of the almost 130% increase in obesity in the United States over the last 40 years (Wang and Beydoun 2007), they constitute an important emerging challenge for EPA science and regulation.

An area of increasing recognition is that of cumulative effects from the built and social environment on health and well-being. Multiple exposures and social factors can interact to increase risks and affect community health status. The role of the built environment in community health is analogous to the role of habitat change in ecologic quality. Effective environmental protection takes into consideration all environments that are valuable to humans and natural systems, and EPA can continue to have significant impacts in this area of research.

Today and in the future, EPA will be challenged to maintain and consider an expanding list of chemicals and potential adverse environmental health effects. Because people are being exposed to many different types of stressors that may interact antagonistically or synergistically and because chemicals can affect different populations in different ways, EPA will also be challenged to refine methods to evaluate cumulative effects (EPA 2011b). New approaches to understanding and managing risks and to measuring health outcomes would support more informed environmental-policy decisions.

Biomonitoring and Emerging Concerns about Exposure and Health

Biomonitoring for human exposure to chemicals in the environment has provided a new lens for understanding population exposures to toxicants. The *Fourth National Report on Human Exposure to Environmental Chemicals* measured 212 chemicals in the US population, including 75 for the first time (CDC 2009). The results indicated some declining loads of historical pollutants, such as lead and polychlorinated biphenyls, but also indicated widespread population exposure to previously unmeasured and potentially toxic chemicals. For example, bisphenol A, which potentially has reproductive and endocrine effects, was found in the urine of over 90% of those sampled. Bioaccumulated polybrominated diphenyl ethers were found in the serum of almost the entire population, as were several polyfluorinated compounds used to impart nonstick characteristics to surfaces. The report also provided improved data on pervasive exposures to historically recognized toxicants, such as arsenic and mercury.

The "exposome" is a measure of all exposures that a person accrues in a lifetime (see Chapter 3). It is exceedingly difficult to measure all exposures that a person accrues in a lifetime because of the enormous variability in exposure over space and time and to an ever-changing set of chemicals that are used by society. Measuring such exposures in an entire population is even more difficult. Yet the exposome is a useful concept that will be increasingly important in coming years and allow the exploration of the progression of disease from an absorbed dose to a targeted health outcome, including the influence of genetic information on susceptibility and biomarkers.

Novel understanding of population exposure brings new challenges for environmental health science. The report *Biomonitoring for Environmental Chemicals* (NRC 2006) indicates the analytic methods for detecting exposures have outpaced the science of interpreting the potential implications for human health. As the list of biomarkers grows, EPA will face constant challenges to interpret health and ecologic implications, identify sources of exposure, and trace the pathways of human exposure. In addition to the traditional single-substance approach, the recognition that the population is chronically exposed to low concentrations of large numbers of pollutants will necessitate new methods for understanding cumulative effects of multiple contaminants on health.

Air Pollution and Climate Change

EPA's first goal in its 2011–2015 strategic plan is "taking action on climate change and improving air quality" (EPA 2010a). This goal encompasses mandates under the Clean Air Act and other statutes, obligations under international treaties and agreements, and executive branch commitments. The following sections provide examples of challenges associated with understanding and addressing air pollution and climate change.

Improving Air Quality

The Clean Air Act is designed primarily to address effects on human health and welfare (including visibility and ecologic effects) that are due to pollutants released into or produced in the ambient atmosphere. That is accomplished through regulations that limit emissions from a broad array of sources—feedlots, ship engines, petroleum refineries, power plants, vehicles, and more. The act requires EPA to protect human health and welfare through provisions that specifically address a core set of six criteria air pollutants, nearly 200 listed hazardous air pollutants, acid deposition, and protection of the stratospheric ozone layer (42 USC [2008]). It also directs the EPA administrator to regulate other air pollutants on finding they may reasonably be expected to endanger public health and welfare. The Clean Air Act and other statutory mandates give rise to the need for improved scientific and technical information on health effects, human exposures, ecologic exposures and effects, ambient and emission monitoring techniques, atmospheric chemistry and physics, and pollution-prevention and emission-control methods for hundreds of pollutants.

Beyond the outdoor air-quality focus under the Clean Air Act, some programs are designed to address indoor air quality. Many Americans spend 65% to over 90% of their time indoors (Allen et al. 2007; Wallace and Ott 2011). Exposures to certain pollutants released from building materials and consumer products are often substantially greater indoors than outdoors (Hoskins 2011). EPA has extensive authority over chemicals and microbial agents found or used in the indoor environment under environmental laws including the Toxic Substances Control Act and the Federal Insecticide, Fungicide, and Rodenticide Act. It also sets the guideline for acceptable levels of radon in indoor air. EPA is a leader in understanding the dynamics of vapor intrusion from soil gas into buildings and it conducts research on human exposure in the indoor environment and corresponding health effects (EPA 2005, 2011c, 2012a,b,c).

The agency's efforts to improve air quality continue to have high priority despite decades of progress because the economic costs that air pollution imposes on society remain high. For example, *Hidden Costs of Energy: Unpriced Consequences of Energy Production and Use* (NRC 2010) estimated that the aggregate damages in the United States associated with air pollution from the country's coal-fired power plants were at least $62 billion in 2005 and that air pollution from motor vehicles contributed at least another $56 billion in damages. The Clean Air Act is an expensive law in terms of compliance, but it still has a highly positive benefit-to-cost ratio (EPA 2011d). EPA recently issued a report called *The Benefits and Costs of the Clean Air Act from 1990 to 2020* (EPA 2011d). According to that study, the direct benefits from the 1990 Clean Air Act amendments are estimated to be almost $2 trillion by the year 2020, exceeding costs by a factor of more than 30 to 1.

Impacts of Climate Change

In the last several decades, it has become clear that human activities have had substantial effects on global climate. The global temperature has increased by an average of 0.6°C since 1901 (IPCC 2007) and variability has increased as well, especially in patterns of precipitation and runoff. That pattern led Milly et al. (2008) to conclude that "stationarity is dead"[1] in the context of water-resource management and to suggest that a new paradigm is needed for dealing with the fact that human society can no longer count on the conformity of mean precipitation—or even variability in annual precipitation—to historic patterns. Many climatologists, while concerned about the increase in mean global temperature, are focused on the changes in extreme temperatures and precipitation (such as floods and droughts) because the extremes cause greater social and ecologic disruption than a shift in average temperatures. Climate change may be the most obvious example of the need for systems thinking in policy-making, given complex interactions between regional air quality and climate change and the numerous pathways by which the environment and human health can be influenced. Many of the factors discussed earlier in this chapter will have direct and indirect influences on climate change, which will itself influence land use patterns and other drivers.

There is evidence that the climate change that has occurred in recent decades has made it harder and more expensive to address air-quality problems (see, for example, Bloomer et al. 2009 and IWGSCC 2010). Furthermore, there is strong scientific consensus that in coming decades climate change is likely to increase the frequency of heat waves, exacerbate problems with water supply and water quality, increase the severity of storms, and disrupt ecosystems, habi-

---

[1]*Stationarity* is the term used when statistics (such as mean, median, variance) are constant through time.

tat, and food production (IPCC 2007, 2012). The scientific and technical challenges associated with the goal of taking action on climate change and improving air quality are broad and complex. Finding efficient and effective approaches to mitigate and adapt to climate change and improve air quality requires systems thinking and research in diverse disciplines, including environmental engineering, atmospheric sciences, biology, ecology, engineering, economics, sociology, and public health. EPA has been involved in climate-change research and policy development for more than 2 decades (see Box 2-1). Beyond its statutory assignments, EPA undertakes broader efforts to address climate change and improve air quality through various approaches that include public education, consumer information, technical exchanges, grants, and voluntary certification programs.

Regulatory Drivers for Air Quality and Climate Change

EPA's regulatory drivers in the climate-change and air-quality arena have helped to marshal resources in and outside the agency, which has yielded substantial advances in scientific understanding and technology. For example, designation of particulate matter and photochemical oxidants as criteria pollutants under the Clean Air Act has led to thousands of epidemiology and toxicology studies that have improved the understanding of associated health effects and provided the scientific basis of standard-setting and regulatory efforts. In contrast, one challenge posed by regulatory drivers is the blind spots that they create for issues deemed outside the scope of regulatory authority or issues that have lower priority because of later deadlines or milder penalties for noncompliance. For example, EPA recognizes both indoor pollution and outdoor air pollution as posing important health risks, but the agency places relatively low priority on indoor air-quality research due to lack of a regulatory mandate. The structure of the Clean Air Act has also encouraged heavier emphasis on criteria pollutants over other hazardous air pollutants, human health over ecosystem effects, and industrial sources over agricultural sources of pollutants. EPA faces a challenge in trying to balance its own research portfolio between issues that arise out of its regulatory mandate and issues that warrant attention from the perspective of human health and welfare but for which there is no legislative mandate. Approaches for how EPA can support and promote science and engineering in the face of these challenges are discussed in Chapters 4 and 5.

Continued research efforts and leadership are important for a strong understanding of the health effects and fate and transport of conventional air pollutants, including both hazardous air pollutants and criteria pollutants, and understanding the synergistic effects of air-pollutant mixtures. EPA would benefit from advancing the understanding of sources, transformations, and transport of pollutants, including improved quantification and forecasting of international contributions to US air-quality challenges (for example, mercury deposition and

> **BOX 2-1** Environmental Protection Agency Involvement in Climate Change
>
> The Global Change Research Act of 1990 established a framework for federal research that continues today as the US Global Change Research Program. EPA is one of 13 agencies and departments participating in the program and has special responsibility for research to assess consequences of global change for air and water quality, aquatic ecosystems, and human health. EPA is responsible for the greenhouse-gas inventory that the United States submits to the secretariat of the United Nations Framework Convention on Climate Change, which the United States ratified in 1992. In 2007, the US Supreme Court held that EPA is responsible for regulating greenhouse gas emissions as air pollutants under the Clean Air Act if the administrator finds that the act's endangerment condition is satisfied. EPA Administrator Lisa Jackson made that finding in December 2009. Accordingly, the agency has set greenhouse-gas emission standards for motor vehicles and is moving forward with greenhouse-gas emission regulations for stationary sources. The Consolidated Appropriations Act of 2008 required EPA to promulgate requirements for large sources of greenhouse-gas emissions to track and report these emissions.

ozone nonattainment). As it grapples with climate change, this type of research would give the agency better understanding of interactions between climate change and air quality with respect to both atmospheric responses and opportunities for mitigation.

Water Quality

During the 1970s, key legislation that focused on developing sound policies for protection of surface water and groundwater was passed, including the Clean Water Act and the Safe Drinking Water Act. Both concentrated on water quality and public health, but the presence of different goals, approaches, and targets led to fragmented water science and research agendas (Table 2-1). It has long been argued that a harmonization of the two acts is needed, and some view a national water quality policy as a threat to or a necessity for achieving secure and safe water supplies and addressing key challenges in the future.

Drivers of Water-Quality Policy

The major drivers for developing national research and science agendas are focused on looming water problems. Since 1970, although understanding of hydrologic systems has advanced, water problems have been overshadowed by the challenges and rapid changes in land use and economic systems (Langpap et al. 2008). Provision of a safe and sustainable supply of water for humanity is widely expected to be one of the central issues of global politics and economics during this century. Water is also closely tied to many other leading sustainability issues such as energy, climate, and food security. Given increasing demands on freshwater supplies, particularly in the more arid regions of the western United States, the challenges of providing clean water are prominent today and will likely continue to be a concern in the future. Demands include domestic uses (potable and landscaping), agricultural uses, and support of ecosystems and biodiversity, and global change will exacerbate the tension among those demands.

The climate–water nexus presents new challenges and will require substantial investment in scientific research for managing this stressed resource in regions where water is scarce and in regions where water is plentiful. Regions experiencing water stress are projected to double by 2050 as a result of climate change (Bates et al. 2008). There is evidence that global climate change will increase the threat to human health, ecosystems, and socioeconomic conditions (IPCC 2007). As previously discussed, there will probably be direct effects on human health due to weather and climate extremes (for example, extremes in temperature and precipitation) and disasters caused by these extreme weather events (such as heat waves, floods, and hurricanes) (IOM 2009). Water is at the heart of understanding climate-change threats, and a new strategy for interdisciplinary research programs is imperative if the threat is to be handled without large adverse effects.

TABLE 2-1 Some Contrasts between the Clean Water Act and the Safe Drinking Water Act

ISSUE	*Clean Water Act of 1972*	*Safe Drinking Water Act of 1976*
Goals	• Swimmable, fishable water • Ecologic quality addressing ambient waters and discharges • Standards developed at the state level	• "Safe" drinking water as defined by maximum contaminant levels for final drinking-water or performance standards • Nationally consistent standards
Technology	• Little advancement in routine wastewater treatment or monitoring • Technologic advances associated with state efforts in wastewater reclamation	• New monitoring tools • New treatment technology to address new contaminants • Sensor technology associated with distribution systems and water security
Science	• Impaired waters and development of hydrologic models • Predictive modeling • Source tracking methods using molecular tools	• Advancement of risk-assessment frameworks and methods • Groundwater models • National databases
Policies	• Beaches Environmental Assessment and Coastal Health Act • Nutrient criteria	• Contaminant Candidate List

The ability to meet the global need for an adequate water supply will come from new scientific insights that span traditional disciplines and from innovative policy based on that science. Global water-research agendas have begun to address needs in the various elements of science, engineering, technology, and policy—drought and flood initiatives associated with climate variability, mitigation of water-related disasters, enhancement of water quality, emerging contaminants, interactions between water and food security, water and human settlements, groundwater sustainability, advanced water-treatment technologies, and ecohydrology. Related cross-cutting issues include the building of research and technology capacity, education, governance, and international relationships associated with water.

In addition to being driven by evolving water-quality problems, water-policy change is likely required to respond to tightened public budgets and increased concerns about efficiency in water-quality regulation (Stoner 2011). For example, water-pollution control in agriculture, a leading cause of non–point-source pollution problems in the United States, has been pursued largely through voluntary compliance strategies supplemented by public assistance through the adoption of pollution control practices. Reduced federal and state budgets may require significant policy innovations if water-quality goals are to be achieved with reduced financial support (Shortle et al. 2011).

Water Technology and Infrastructure Research

Monitoring technology is a vital component of water science. Emerging concerns about contaminants have appeared dramatically (for example, the outbreak of *Cryptosporidium* in Milwaukee, Wisconsin; Mac Kenzie et al. 1994) and resulted in the need for tools to be developed quickly or have arisen via advances in analytic capabilities (for example, identification of pharmaceuticals in the water supply). (See Chapter 3 for a discussion of these tools.) Although the health effects of some contaminants are clear, in most cases there are a host of reasons why the Clean Water Act and the Safe Drinking Water Act have resulted in a limited record of accomplishments. Some of those reasons include, low concentrations found in water, specific limitations of the methods for pathogen recovery and viability assessment, failure to understand whether ingestion or inhalation pathways are important, and inability to reconcile ecologic risks and human health risks. The inadequate investment in scientific inquiry associated with sources, transport, and fate of contaminants has led to much uncertainty about the most effective risk reduction management approaches.

Advances in other fields have had important impacts on water science. Nanomaterials, discussed further in Chapters 3 and 4, are a case in point. Although nanomaterials have the opportunity to support novel water-treatment approaches and more efficient disinfection, there is heightened concern about nanoparticles as a contaminant and about the inability to measure and monitor their fate. Nonetheless, nanomaterials may play a role in "tunable" reactive membranes for desalination, water reuse, and disinfection in the future (Savage and Diallo 2005; Wiesner and Bottero 2007). Various nanostructured catalytic membranes could be used to selectively kill pathogens in drinking water, remove ultratrace contaminants from wastewater for water reuse, or provide bioactive degradation of pharmaceuticals or hormonally active substances from drinking water.

A very basic problem in the near future is how to replace existing, aging infrastructure in the face of a growing population and declining resources. Much of US water and wastewater infrastructure is nearly 100 years old and in dire need of modernization and replacement. The American Society of Civil Engineers grades the US water and wastewater infrastructure as "D-". It estimates the 5-year investment needed for America's infrastructure is over $2.2 trillion dollars (ASCE 2009). In some cases, the best designs for replacing current infrastructure may be radically different from the past (decentralized vs. centralized; large built structures vs. small green infrastructure; and low impact development, water reuse, or desalinization). Science and engineering research, coupled with systems-thinking approaches that take account of the numerous implications of water infrastructure, will determine the most cost-effective processes and infrastructure.

Nutrient Pollution

Nitrogen and phosphorus are essential nutrients that control the growth of plants and animals. However, problems occur when excess inputs cause large increases in aquatic plant and algal growth and in turn changes in plant and algal species (Bushaw-Newton and Sellner 1999). Decaying algal blooms consume dissolved oxygen, and this leads to hypoxic conditions that are harmful or deadly for many aquatic organisms. Nutrient pollution can cause important economic losses through damage to commercial and recreational fisheries, restrictions on contact-based water recreation, and disamenities (EPA 2012d). Nitrates also pose a human health risk when present at high concentrations in drinking water.

Water-quality conditions reported by states under the Clean Water Act indicate that at least 100,000 miles of rivers and streams; nearly 2.5 million acres of lakes, reservoirs, and ponds; and over 800 square miles of bays and estuaries across the United States are listed as impaired and not meeting state water-quality goals as a result of nitrogen and phosphorus enrichment (EPA 2012a). Only a small fraction of the nation's total water resources are currently assessed, so those values are underestimates of the spatial extent of nutrient-impaired waters (EPA 2006, 2010a). Diaz and Rosenberg (2008) found that dead zones in the coastal oceans of the world have increased exponentially since the 1960s, and many of them are located along the US Atlantic and Gulf of Mexico coasts. Harmful algal blooms have been reported in virtually all US coastal waters (Bushaw-Newton and Sellner 1999), and symptoms of eutrophication have been found in 78% of the assessed continental US coastal area (Selman et al. 2008).

Excess nutrients reach surface-water resources in direct discharges from point sources (for example, municipal wastewater-treatment plants) and from diffuse non–point sources (for example, nutrient runoff from farmland, urban, and suburban areas and air pollution). Because the nutrient-use efficiency of crops is less than 100%, farmers need to apply more nutrients to their fields than the plants need for healthy growth. The challenge for all farmers is to add fertilizer at the optimal time and rate and then to keep the nutrients in the field. Concomitant with the substantial increases in agronomic yields that have allowed agriculture and fish production to meet the food needs of 7 billion people has been a need for higher rates of application of fertilizers, which have exacerbated runoff, limited the effectiveness of strategies for remediating eutrophication, and resulted in production of nitrous oxide as a byproduct of nitrification and denitrification processes. (Nutrient sources for the Chesapeake Bay and the Gulf of Mexico are shown in Figure 2-1.) Addressing the nutrient loading will require increased scientific understanding, including new information on pollution sources, on emerging technologies that could be used in agriculture and in wastewater treatment, on water quality conditions, and on the response of ecosystems to increasing nutrient loads and shifting stochiometry. Such scientific understanding can be gained only through integrated research.

The Chesapeake Bay, North America's largest estuary, offers a highly instructive example of contributions made by EPA and allied researchers to a more fundamental understanding of the physical processes that lead to the effects of nutrient pollution. Substantial reductions in nutrient discharges from sewage-treatment plants, factories, and other point sources of pollution have been achieved in the bay watershed since the 1970s but are insufficient to meet water-quality goals. The challenges faced by the Chesapeake Bay ecosystem are shared by many other ecosystems, but the differences among them make the required research and the effective tools for addressing the challenges more complex. For example, 500 km to the north of the Chesapeake Bay lies Narragansett Bay. Although smaller than its southern cousin, it shares many historical and ecologic characteristics; but the challenges faced today by the Narragansett Bay (where EPA's Atlantic Ecology Division Laboratory is located) have developed in very different ways. The region has historically been dominated by agricultural activity, but that is no longer the case. Today, Narragansett Bay suffers from excess nitrogen inputs, largely from upstream wastewater-treatment facilities (Pryor et al. 2007). The upper reaches of the bay have been closed to shellfishing and swimming for decades. In 2004, Rhode Island mandated a minimum standard for effluent nitrogen from the wastewater facilities within its jurisdiction, yet the science suggests that without concomitant reductions in nitrogen from wastewater facilities upstream on the Blackstone River in Massachusetts and reduction in nitrogen inputs that result directly and indirectly from air pollution, restoring the waters of the upper bay will be difficult (see Figure 2-2). Narragansett Bay, as a result of the large influence of sewered effluents, should be one of the easiest places to address chronic water-quality deterioration, but it has proved elusive even there.

Nutrient (nitrogen and phosphorus) pollution is one of the more persistent and pervasive environmental problems in the United States, and it is worsening in many locations (Howarth 2008). The volume of nutrients reaching surface water and groundwater has increased substantially since the middle of the 20th century as a result of a complex of factors, including population growth, changes in land cover, increased fossil fuel combustion, and changes in the structure of agricultural production (Selman et al. 2008). Providing the scientific foundations for the development of policies that can reduce nutrient-pollution problems will require innovative economic, social-science, and natural-science research. The challenges are particularly difficult because the hydrologic, ecologic, economic, and social processes affecting the magnitude and scope of nutrient pollution and its consequences are complex, multi-scaled, and spatially variable. To deal effectively with this complex problem, a framework for incorporating human and environmental interactions, such as the Millennium Ecosystem Assessment framework (see Chapter 1) would prove useful. Nutrient pollution should be approached from a broad perspective that uses systems thinking (see Chapter 4) and there are examples in which EPA is already taking steps in this direction with the Chesapeake Bay Program and the New York–New Jersey Harbor Estuary Program. The problem may not be getting progressively worse, but there are still many challenges to attaining further improvements. The prospects are that eutrophication will continue to be a challenge until policies to control nutrients are made more effective (Cary and Migliaccio 2009; Spiertz 2009).

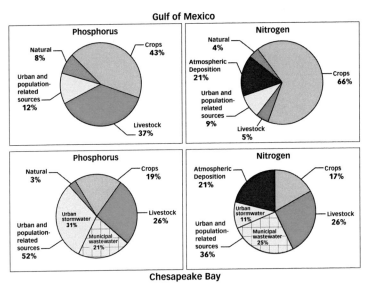

FIGURE 2-1 Sources of phosphorus and nitrogen in the Gulf of Mexico and Chesapeake Bay. Source: EPA 2010b.

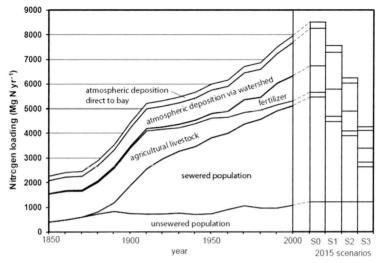

FIGURE 2-2 Narragansett Bay nitrogen loading from 1850 to 2015 under several different scenarios. Scenario 0 (S0), current conditions and no improvements in wastewater treatment; scenario 1 (S1), current conditions and implementation of all mandated reductions in nutrient from wastewater-treatment plants; scenario 2 (S2), all wastewater treatment plants have a maximum effluent nitrogen of 8 mg/L in summer, 25% reduction in nitrogen air-pollution concentrations, and 25% reduction in fertilizer use in the watershed; scenario 3 (S3), all wastewater-treatment plants have a maximum effluent nitrogen concentration of 3mg/L in summer, 50% reduction in nitrogen air-pollution concetrations, and 50% reduction in fertilizer use in the watershed. Source: Vadeboncouer et al. 2010. Reprinted with permission; copyright 2010, *Estuaries and Coasts.*

Shifting Spatial and Temporal Scales

In the early days of environmental remediation and pollution control, the problems were more obvious. One could see, indeed often even smell or taste, the pollutants, and local causes could be easily identified. As progress has been made in cleaning up the local problems and as more has been learned about the health and environmental consequences of chronic low-dose exposures to diverse chemicals, much of the focus has moved to wider geographic areas. The spatial scales required to understand emerging environmental issues vary widely and are increasing as more is learned about the systems underlying the observed phenomena.

Acid rain and photochemical air pollution are regional problems, and monitoring, modeling, and control activities have shifted accordingly. EPA's long-standing involvement in regional-scale air quality monitoring and modeling research includes the multi-agency National Acid Precipitation Assessment Program (NAPAP 1991), which was authorized by Congress in 1980 and informed the acid rain provisions of the 1990 Clean Air Act amendments. EPA continues to work with other federal and state agencies to improve understanding of the nature and consequences of air pollutant deposition to terrestrial and aquatic ecosystems on regional scales. More recently, EPA has conducted and supported research linking global climate projections to regional-scale air quality (EPA 2009), which has demonstrated the potential for global climate change to exacerbate the challenge of meeting health-based air quality standards. Regional long-term approaches for assessment and problem-solving have also been implemented in the water quality arena, including for the Chesapeake Bay, the Florida Everglades, and the Great Lakes Basin (Table 2-2). In the future, EPA will need to develop a better understanding of the sources, transport, and fate of global-scale pollutants to avoid the possibility that little improvement in environmental quality occurs even when local investment is large. For example, although lead from local sources, such as coal-fired power plants, is important, these local emissions are superimposed on a global background of lead, some of which is transported on intercontinental scales from both natural and anthropogenic sources (UNEP 2006). Mercury transport at the regional and global scale is another example. It is not feasible for EPA to undertake all the global-scale monitoring and modeling that are needed, but it can work proactively with other US federal agencies (such as the National Oceanic and Atmospheric Administration, the National Aeronautics and Space Administration, and the National Science Foundation) and with international organizations to ensure that the issues that it most needs to understand remain high on research agendas. (See Chapter 4 for a discussion on collaboration.) Current environmental challenges are expanding not only in space but also in time. Some responses to perturbations are rapid (such as algal blooms), others are slow (such as vegetation response to climate change). To understand how and why these effects unfold, long-term data are needed to characterize the changes, the causes, and the potential implications of different policy options. (The needs for such data are discussed further in Chapters 3 and 4.) Without the perspective provided by long-term data, it is easy to assume wrongly that short-term variations in environmental characteristics reflect substantive changes in the environment, and it is easy to miss important but subtle or slow changes in the environment.

SUMMARY

This chapter discusses some of the major factors driving changes in the environment and gives illustrative examples of the complex and multidisciplinary challenges that EPA faces now and will probably face in the future. To address those challenges, EPA will need to continue to develop and support scientific methods, tools, and technologies that apply a systems-thinking approach to understand environmental changes and their effects on human health (see Chapter 4).

TABLE 2-2 Large Regional Water Programs in the US Environmental Protection Agency

Watershed or Water System	Key Stressors and Issues	EPA Leadership	Science and Engineering Focus
Chesapeake Bay, North America's largest estuary (EPA 2012e)	• Eutrophication caused by nutrient enrichment • Nitrogen • Stressed by pressures of growing populations, industrial pollution, atmospheric deposition of air pollutants, and conversion of forests to farms and urban areas	• EPA, Region 3 Mid-Atlantic • Chesapeake Bay: A Framework for Action (EPA 1983a,b) provided a framework for additional research and policy initiatives based on the Chesapeake Bay Program, which was established in 1983 as a partnership of the EPA, Maryland, and Virginia	• Developed a number of innovative tools • Nitrogen-removal technology at wastewater-treatment plants • Computer model built to simulate how the massive 64,000-mi^2 watershed processes nutrient and sediment allocations
Everglades, a sub-tropical wetlands watershed Florida, which houses the Everglades National Park (EPA 2007)	• Altered hydrology • Mercury • Phosphorus • Soil erosion • One of the most threatened subtropical preserves in the United States	• EPA, Region 4 • The Comprehensive Everglades Restoration Plan is an ambitious, multi-billion-dollar and multi-decadal restoration program involving federal and state governments • EPA developed the Everglades Ecosystem Assessment Program, which contributes to the joint federal–state Comprehensive Everglades Restoration Plan	• Aquifer storage and recovery • Ecosystem restoration • One of the strongest aspects of the Comprehensive Everglades Restoration Plan science program is its monitoring and assessment program (see, for example, NRC 2003) • Developed scientific tests, experiments, and physical models
Great Lakes Basin, the largest transboundary freshwater system in the world (Lakes Erie, Huron, Michigan, Ontario, and Superior and five major connecting rivers: Detroit, Niagara, St. Clair, St. Lawrence, and St. Mary's) (EPA 2011e).	• Climate change associated with lowering lake level • Invasive species • Nutrients • Pathogens • Mercury loading (alone and with contaminated sediments) • Effects on community health, tourism, fisheries, power industry, and grids; human health and ecosystems are seen as being at risk	• EPA, Region 5 • Great Lakes Restoration Initiative action plan • A Great Lakes inter-agency task force was formed to coordinate federal and bi-national restoration efforts	• The largest investment in the Great Lakes in two decades • Priority "focus areas" were "1) cleaning up toxics and toxic hot spot areas of concern; 2) combating invasive species; 3) promoting near-shore health by protecting watersheds from polluted run-off; 4) restoring wetlands and other habitats; and 5) tracking progress, education, and working with strategic partners" (MI DNR 2011)

The drivers outlined in this chapter are often overlapping and their nature is changing over time. For example, in the United States, chemical exposures from industrial facilities are decreasing significantly; dispersed, non-point, and less controllable exposures from chemicals used in products may represent a larger percentage of the current chemical burden to ecosystems and humans. As illustrated by the degradation of the Chesapeake Bay, multiple overlapping factors, such as land use and changing land-use patterns, population growth, the agricultural use of fertilizers and pesticides, and direct and non-point chemical exposures may result in human and environmental effects. The complexity of these interacting factors in environmental degradation creates great challenges for environmental science and decision-making.

The siloed, disciplinary approaches that have often been taken to monitor for and characterize singular types of effects and to develop control measures will not be sufficient to understand and prevent environmental changes and their health effects. There is a need for greater attention to understand the complex systems in which human activities are causing effects and how those effects interact. Ultimately, prevention of these complex effects will require greater systematic efforts to understand the way in which products, consumptive systems (such as energy), communities, and other human activities are designed and carried out.

REFERENCES

Allen, J.G., M.D. McClean, H.M. Stapleton, J.W. Nelson, and T.F. Webster. 2007. Personal exposure to polybrominated diphenyl ethers (PBDEs) in residential indoor air. Environ. Sci. Technol. 41(13):4574-4579.

Alvarez, R.A., S.W. Pacala, J.J. Winebrake, W.K. Chameides, and S.P. Hamburg. 2012. Greater focus needed on methane leakage from natural gas infrastructure. Proc. Natl. Acad. Sci. USA 109(17):6435-6440.

ASCE (American Society of Civil Engineers). 2009. Report Card 2009 Grades [online]. Available: https://apps.asce.org/reportcard/2009/grades.cfm [accessed April 29, 2012].

Bates, B.C., Z.W. Kundzewicz, S.Wu, and J.P. Palutikof, eds. 2008. Climate Change and Water. IPCC Technical Paper VI. Geneva: Intergovernmental Panel on Climate Change [online]. Available: http://www.ipcc.ch/pdf/technical-papers/climate-change-water-en.pdf [accessed Mar. 28, 2012].

Bloomer, B.J., J.W. Stehr, C.A. Piety, R.J. Salawitch, and R.R. Dickerson. 2009. Observed relationship of ozone air pollution with temperature and emissions. Geophys. Res. Lett. 36:L09803.

Brownstone, D. and T.F. Golob. 2009. The impact of residential density on vehicle usage and energy consumption. Journal of Urban Economics. 65:91-98.

Bullard, R.D. 2000. Dumping in Dixie: Race, Class, and Environmental Quality, 3rd Ed. Boulder, CO: Westview Press.

Bushaw-Newton, K.L., and K.G. Sellner. 1999. Harmful Algal Blooms. In NOAA's State of the Coast Report. Silver Spring, MD: National Oceanic and Atmospheric Administration [online]. Available: http://oceanservice.noaa.gov/websites/retiredsites/sotc_pdf/hab.pdf [accessed Mar. 27, 2012].

Cary, R.O., and K.W. Migliaccio. 2009. Contribution of wastewater treatment plant effluents to nutrient dynamics in aquatic systems: A review. Environ. Manage. 44(2): 205-217.
CDC (Centers for Disease Control and Prevention). 2009. Fourth National Report on Human Exposure to Environmental Chemicals. Department of Health and Human Services, Centers for Disease Control and Prevention [online]. Available: http://www.cdc.gov/exposurereport/pdf/FourthReport.pdf [accessed Mar. 27, 2012].
Confalonieri, U., B. Menne, R. Akhatar, K.L. Ebi, M. Haugengue, R.S. Kovats, B. Revich, and A. Woodward. 2007. Human health. Pp. 391-431 in Climate Change 20007: Impacts, Adaptation and Vlunterability. Contribution of Working Group II to the Fourth Assessment Report of the Intergovernmental Panel on Climate Change, M.L. Parry, O.F. Canziani, J.P. Palutikof, P.J. van der Linden, and C.E. Hanson, eds. Cambridge: Cambridge University Press [online]. Available: http://www.ipcc.ch/pdf/assessment-report/ar4/wg2/ar4-wg2-chapter8.pdf [accessed Apr. 9, 2012].
Diaz, R.J., and R. Rosenberg. 2008. Spreading dead zones and consequences for marine ecosystems. Science 321(5891):926-929.
EPA (US Environmental Protection Agency). 1983a. Chesapeake Bay: A Framework for Action. US Environmental Protection Agency, Region 3, Philadelphia, PA [online]. Available: http://www.chesapeakebay.net/content/publications/cbp_12405.pdf [accessed Mar. 29, 2012].
EPA (US Environmental Protection Agency). 1983b. Chesapeake Bay: A Framework for Action-Appendices. US Environmental Protection Agency, Region 3, Philadelphia, PA [online]. Available: http://www.chesapeakebay.net/content/publications/cbp_132 62.pdf [accessed Mar. 29, 2012].
EPA (US Environmental Protection Agency). 2005. Program Needs for Indoor Environments Research (PNIER). EPA 402-B-05-001. US Environmental Protection Agency, March 2005 [online]. Available: http://www.epa.gov/iaq/pdfs/pnier.pdf [accessed July 20, 2012].
EPA (US Environmental Protection Agency). 2006. Wadeable Streams Assessment: A Collaborative Survey of the Nation's Streams. EPA 841-B-06-002. Office of Water, US Environmental Protection Agency, Washington, DC [online]. Available: http://www.cpcb.ku.edu/datalibrary/assets/library/projectreports/WSAEPArepo rt.pdf [accessed Mar. 28, 2012].
EPA (US Environmental Protection Agency). 2007. Everglades Ecosystem Assessment: Water Management and Quality, Eutrophication, Mercury Contamination, Soils and Habitat-Monitoring for Adaptive Management. EPA 904-R-07-001. US Environmental Protection Agency, Region 4, Athens, GA [online]. Available: http://www.epa.gov/region4/sesd/reports/epa904r07001.html [accessed Mar. 29, 2012].
EPA (US Environmental Protection Agency). 2009. Assessment of the Impacts of Global Change on Regional US Air Quality: A Synthesis of Climate Change Impacts on Ground-Level Ozone, An Interim Report of the US EPA Global Change Research Program. EPA/600/R-07/094F. National Center for Environmental Assessment, Office of Research and Development, US Environmental Protection Agency, Washington, DC [online]. Available: http://www.mwcog.org/uploads/committee-documents/bV5cWVxb20090417102640.pdf [assessed July 31, 2012].
EPA (US Environmental Protection Agency). 2010a. FY 2011-2015 EPA Strategic Plan: Achiving Our Vision. US Environmental Protection Agency, September 30, 2010 [online]. Available: http://www.epa.gov/planandbudget/strategicplan.html [accessed Jan. 17, 2012].

EPA (US Environmental Protection Agency). 2010b. Discussion Document: Coming Together for Clean Water, Background Information on Proposed Discussion Topics. US Environmental Protection Agency Forum, April 15, 2010 [online]. Available: http://blog.epa.gov/waterforum/discussion-document/ [accessed Mar. 29, 2012].

EPA (US Environmental Protection Agency). 2011a. Inventory of US Greenhouse Gas Emissions and Sinks: 1990-2009. EPA 430-R-11-005. US Environmental Protection Agency, Washington, DC. April 15, 2011[online]. Available: http://www.epa.gov/climatechange/emissions/downloads11/US-GHG-Inventory-2011-Complete_Report.pdf [accessed Mar. 28, 2012].

EPA (US Environmental Protection Agency). 2011b. Human Disease and Condition. Report on the Environment [online]. Available: http://cfpub.epa.gov/eroe/index.cfm?fuseaction=list.listBySubTopic&ch=49&s=381 [accessed Mar. 6, 2012].

EPA (US Environmental Protection Agency). 2011c. Exposure Model for Individuals (EMI). Human Exposure and Atmospheric Science Program, US Environmental Protection Agency [online]. Available: http://www.epa.gov/heasd/products/emi/emi.html [accessed July 20, 2012].

EPA (US Environmental Protection Agency). 2011d. The Benefits and Costs of the Clean Air Act from 1990 to 2020. Office of Air and Radiation, US Environmental Protection Agency, March 2011 [online]. Available: http://www.epa.gov/oar/sect812/feb11/fullreport.pdf [accessed July 22, 2012].

EPA (US Environmental Protection Agency). 2011e. Great Lakes. US Environmental Protection Agency [online]. Available: http://epa.gov/greatlakes/ [accessed Mar. 29, 2012].

EPA (US Environmental Protection Agency). 2012a. An Introduction to Indoor Air Quality (IAQ). [online]. Available: http://www.epa.gov/iaq/ia-intro.html#content [accessed July 20, 2012].

EPA (US Environmental Protection Agency). 2012b. Gas and Vapor Intrusion. Ground Water and Ecosystems Restoration Reseach, US Environmental Protection Agency [online]. Available: http://www.epa.gov/ada/gw/vapor.html [accessed July 20, 2012].

EPA (US Environmental Protection Agency). 2012c. Soil Gas Sample Collection and Sample Analysis Methods. Underground Storage Tanks, Office of Solid Waste and Emergency Response, US Environmental Protection Agency [online]. Available: http://www.epa.gov/oust/cat/pvi/soil_gas.htm [accessed July 20, 2012].

EPA (US Environmental Protection Agency). 2012d. Nutrient Pollution [online]. Available: http://epa.gov/nutrientpollution/index.html [accessed Mar. 28, 2012].

EPA (US Environmental Protection Agency). 2012e. Chesapeake Bay Program Office. US Environmental Protection Agency Region 3: The Mid-Atlantic States [online]. Available: http://www.epa.gov/region3/chesapeake/ [accessed Mar. 29, 2012].

Erdmann, L., and T.E. Graedel. 2011. Criticality of non-fuel minerals: A review of major approaches and analyses. Environ. Sci. Technol. 45(18):7620-7630.

Frumkin, H. 2002. Urban sprawl and public health. Public Health Rep. 117(3):201-217.

Grun, F., and B. Blumberg. 2006. Environmental obesogens: Organotins and endocrine disruption via nuclear receptor signalling. Endocrinology 147(6 suppl.): S50-S55.

Holloway, A.C., D.Q. Cuu, K.M. Morrison, H.C. Gerstein, and M.A. Tarnopolsky. 2007. Transgenerational effects of fetal and neonatal exposure to nicotine. Endocrine 31(3):254-259.

Hoskins, J.A. 2011. Health effects due to indoor air pollution. Pp. 665-676 in Survival and Sustainability: Environmental Concerns in the 21st Century, H. Gökçekus, U. Türker, and J.W. LaMoreaux, eds. Environmental Earth Sciences 5. Berlin: Springer.

Howarth, R.W. 2008. Coastal nitrogen pollution: A review of sources and trends globally and regionally. Harmful Algae 8(1):14-20.

IOM (Institute of Medicine). 2009. Global Issues in Water, Sanitation, and Health: Workshop Summary. Washington, DC: National Academies Press.

IPCC (Intergovernmental Panel on Climate Change). 2007. Summary for policy makers. Pp. 1-19 in Climate Change 2007: The Physical Science Basis, S. Solomon, M. Manning, Z. Chen, M. Marquis, K.B. Averyt, M. Tignor, and H.L. Miller, eds. Cambridge, UK: Cambridge University Press.

IPCC(Intergovernmental Panel on Climate Change). 2012. The summary for policymakers. Pp. 1-19 in Managing the Risks of Extreme Events and Disasters to Advance Climate Change Adaptation, C.B. Field, V. Barros, T.F. Stocker, D. Qin, D.J. Dokken, K.L. Ebi, M.D. Mastrandrea, K.J. Mach, G.K. Plattner, S.K. Allen, M. Tignor, and P.M. Midgley, eds. A Special Report of Working Groups I and II of the Intergovernmental Panel on Climate Change. Cambridge, UK: Cambridge University Press [online]. Available: http://ipcc-wg2.gov/SREX/images/uploads/SREX-SPM_FINAL.pdf [accessed Mar. 28, 2012].

IWGSCC(Interagency Working Group on Social Cost of Carbon). 2010. Social Cost of Carbon for Regulatory Impact Analysis – Under Executive Order 12866. Technical Support Document. Interagency Working Group on Social Cost of Carbon, US Government [online]. Available: http://www.epa.gov/oms/climate/regulations/scc-tsd.pdf [accessed July 21, 2012].

Jacob, D.J., and D.A. Winner. 2009. Effect of climate change on air quality. Atmospheric Environment. 43:51-63.

Jaramillo, P., W.M. Griffin, and S. Matthews. 2007. Comparative life-cycle air emissions of coal, domestic natural gas, LNG, and SNG for electricity generation. Environ. Sci. Technol. 41(17):6290-6296.

Kirchner, S., T. Kieu, C. Chow, S. Casey, and B. Blumberg. 2010. Prenatal exposure to the environmental obesogen tributyltin predisposes multipotent stem cells to become adipocydes. Mol. Endocrinol. 24(3):526-539.

Langpap, C. I. Hascic, and J. Wu. 2008. Protecting watershed ecosystems through targeted local land use policies. Am. J. Agr. Econ. 90(3):684-700.

Mac Kenzie, W.R., N.J. Hoxie, M.E. Proctor, M.S. Gradus, K.A. Blair, D.E. Peterson, J.J. Kazmierczak, D.G. Addiss, K.R. Fox, J.B. Rose, and J.P. Davis. 1994. A massive outbreak in Milwaukee of Cryptosporidium infection transmitted through the public water supply. N. Engl. J. Med. 331(3):161-167.

MI DNR (Michigan Department of Natural Resources). 2011. Michigan Benefits from Great Lakes Restoration Initiative Grants: Over $1.5 Million in Grants Awarded to Local Communities and Organizations [online]. Available: http://www.michigan.gov/dnr/0,4570,7-153--265166--RSS,00.html [accessed Mar. 29, 2012].

Milly, P.C.D., J Betancourt, M. Falkenmark , R.M. Hirsch, Z.W. Kundzewicz, D.P. Lettenmaier, and R.J. Stouffer. 2008. Stationarity is dead: Whither water management. Science 319(5863):573-574.

NAPAP (US National Acid Precipitation Assessment Program). 1991. 1990 Integrated Assessment Report. Washington DC: NAPAP Office of the Director. November 1991.

NOAA (National Oceanic and Atmospheric Administration). 2012a. Keeping Seafood Safe. National Oceanic and Atmospheric Administration [online]. Available: http://www.noaa.gov/100days/Keeping_Seafood_Safe.html [accessed Feb. 27, 2012].

NOAA (National Oceanic and Atmospheric Administration). 2012b. State of the Coast. National Oceanic and Atmospheric Administration [online]. Available: http://stateofthecoast.noaa.gov/ [accessed Mar. 28, 2012].

Nowak, D.J., and E.J. Greenfield. 2012. Tree and impervious cover in the United States. LandscapeUrban Plan. 107(1):21-30.

NRC (National Research Council). 2003. Adaptive Monitoring and Assessment for the Comprehensive Everglades Restoration Plan. Washington, DC: National Academies Press.

NRC (National Research Council). 2006. Human Biomonitoring for Environmental Chemicals. Washington, DC: National Academies Press.

NRC (National Research Council). 2008. Water Implications of Increased Biofuel Production in the US Washington DC: National Academies Press.

NRC (National Research Council). 2010. Hidden Costs of Energy: Unpriced Consequences of Energy Production and Use. Washington, DC: National Academies Press.

NRC (National Research Council). 2011. Renewable Fuel Standard: Potential Economic and Environmental Effects of US Biofuel Policy. Washington DC: National Academies Press.

Pejchar, L., and H.A. Mooney. 2009. Invasive species, ecosystem services and human well-being. Trends Ecol. Evol. 24(9):497-504.

Pimentel, D., R. Zuniga, and D. Morrison. 2005. Update on the environmental and economic costs associated with alein-invasive species in the United States. Ecological Economics. 52(3):273-288.

Pryor, D., E. Saarman, D. Murray, and W. Prell. 2007. Nitrogen Loading from Wastewater Treatment Plants to Upper Narragansett Bay. Narragansett Bay Collection Paper 2. University of Rhode Island [online]. Available: http://digitalcommons.uri.edu/cgi/viewcontent.cgi?article=1002&context=nbcollection [accessed Mar. 28, 2012].

Savage. N.. and M.S. Diallo. 2005. Nanomaterials and water purification: Opportunities and challenges. J. Nanopart. Res. 7(4-5):331-342.

Schmitz, O.J., and T.E. Graedel. 2010. The Consumption Conundrum: Driving the Destruction Abroad. Environment 360, April 26, 2010 [online]. Available: http://e360.yale.edu/feature/the_consumption_conundrum_driving_the_destruction_abroad/2266 [accessed Mar. 28, 2012].

Selman, M.S., S. Greenhalgh, R. Diaz, and Z. Sugg. 2008. Eutrophication and Hypoxia in Coastal Areas: A Global Assessment of the State of Knowledge. WRI Policy Note No. 1. World Resources Institute [online]. Available: http://pdf.wri.org/eutrophication_and_hypoxia_in_coastal_areas.pdf [accessed Mar. 28, 2012].

Shortle, J.S., M. Ribaudo, R.D. Horan, and D. Blanford. 2011. Reforming agricultural nonpoint pollution policy in an increasingly budget-constrained environment. Environ. Sci. Technol. 46(3):1316-1325.

Spiertz, J.H.J. 2009. Nitrogen, sustainable agriculture and food security: A review. Pp. 635-651 in Sustainable Agriculture, E. Lichtfouse, M. Navarrete, P. Debaeke, V. Souchere, and C. Alberola, eds. Dordrecht: Springer.

Squires, G.D. 2002. Urban Sprawl: Causes, Consequences and Policy Responses. Washington, DC: The Urban Institute.

Stoner, N.K. 2011. Testimony of Nancy K. Stoner, Acting Assistant Administrator for Water, US Environmental Protection Agency Before the Subcommittee on Water Resources and Environment Committee on Transportation and Infrastructure US

House of Representatives, June 24, 2011 [online]. Available: http://republicans.transportation.house.gov/Media/file/TestimonyWater/2011-12-14-Stoner.pdf [accessed Mar. 28, 2012].

UN (United Nations). 1999. The World at Six Billion. ESA/P/WP.154. Population Division, Department of Economic and Social Affairs, UN Secretariat [online]. Available: http://www.un.org/esa/population/publications/sixbillion/sixbillion.htm [accessed July 31, 2012].

UN (United Nations). 2011. World Population to Reach 7 Billion on 31 October. United Nations Population Fund Press Release: May 3, 2011 [online]. Available: http://www.unfpa.org/public/home/news/pid/7597 [accessed Mar. 27, 2012].

UNEP (United Nations Environmental Programme). 2006. Report of the First Meeting of the Lead and Cadmium Working Group, September 18-22, 2006, Geneva, Switzerland. United Nations Environmental Programme [online]. Available: http://www.unep.org/hazardoussubstances/Portals/9/Lead_Cadmium/docs/Interim_reviews/K0653416%20-%20Lead%20Cadmium%20Report.pdf [accessed Mar. 28, 2012].

US Census Bureau. 2012. World POPClock Projection [online]. Available: http://www.census.gov/population/popclockworld.html [accessed Mar. 28, 2012].

USDA (US Department of Agriculture). 2006. The human influenced forest. Pp. 19-28 in The State of Chesapeake Forests, E. Sprague, D. Burke, S. Claggett, and A. Todd, eds. Arlington, VA: Conservation Fund [online]. Available: http://www.na.fs.fed.us/watershed/pdf/socf/Full%20Report.pdf [accessed July 6, 2012].

USDA (US Department of Agriculture). 2012. Gateway to Invasive Species Information Covering Federal, State, Local, and International Sources. US Department of Agriculture, National Invasive Species Information Center (NISIC) [online]. Available: http://www.invasivespeciesinfo.gov [accessed Mar. 28, 2012].

Vadeboncouer, M.A., S.P. Hamburg, and D. Pryor. 2010. Modeled nitrogen loading to Narragansett Bay: 1850 to 2015. Estuar. Coast. 33(5):1113-1137.

Wallace, L., and W. Ott. 2011. Personal exposure to ultrafine particles. J. Expo. Sci. Environ. Epidemiol. 21(1):20-30.

Wang, Y., and M.A. Beydoun. 2007. The obesity epidemic in the United States—gender, age, socioeconomic, racial/ethnic, and geographic characteristics: A systematic review and meta-regression analysis. Epidemiol. Rev. 29(1):6-28.

Wiesner M.R., and J-Y. Bottero. 2007. Environmental Nanotechnology: Applications and Impacts of Nanomaterials. New York: McGraw Hill. 540 pp.

Worldwatch Institute. 2011. The State of Consumption Today. Worldwatch Institute [online]. Available: http://www.worldwatch.org/node/810#1 [accessed Feb. 27, 2012].

Wulder, M.A., J.G. Masek, W.B. Cohen, T.R. Loveland, and C.E. Woodcock. 2012. Opening the archive: How free data has enabled the science and monitoring promise of Landsat. Remote Sens. Environ. 122:2-10.

Younger, M., H.R. Morrow-Almeida, S.M. Vindigni, and A.L. Dannenberg. 2008. The built environment, climate change, and health opportunities for co-benefits. Am. J. Prev. Med. 35(5):517-526.

3

Using Emerging Science and Technologies to Address Persistent and Future Environmental Challenges

Chapter 2 discussed some of the broad drivers and challenges that are inherent to the mission of the US Environmental Protection Agency (EPA) today and in the future. Remarkable progress has been made in the last several decades in the development of new scientific approaches, tools, and technologies relevant to addressing those challenges. The purpose of this chapter is to highlight new and changing science and technologies that are or will be increasingly important for science-informed policy and regulation in EPA.

New tools and technologies can substantially improve the scientific basis of environmental policy and regulations, but it is important to remember that many of the tools and technologies need to build on and enhance the current foundation of environmental science and engineering in the United States. In addition, addressing the complex "wicked problems" facing EPA today and in the future requires not only new science and technology but a more deliberate approach to systems thinking, for example, by using frameworks that strive to integrate a broader array of interactions between humans and the environment. From the perspective of scientific advances relevant to the future of EPA, it will be increasingly important that all aspects of biologic sciences and environmental sciences and engineering—including human health risk assessment, microbial pathogenesis, ecosystem energy and matter transfers, and ecologic adaptation to climate change—be considered in an integrated systems-biology approach. That approach must also be integrated with considerations of environmental, social, behavioral, and economic impacts.

A SIMPLE PARADIGM FOR DATA-DRIVEN, SCIENCE-INFORMED DECISIONS IN THE ENVIRONMENTAL PROTECTION AGENCY

New scientific advances, including the development and application of new tools and technologies, are critical for the science mission of EPA. Effective science-informed regulation and policy aimed at protecting human health and environmental quality relies on robust approaches to data acquisition and to knowledge generated from the data. For science to inform regulation and policy effectively, a strong problem-formulation step is needed. Once a problem is formulated, EPA scientists can evaluate what types of data are needed and then determine which available tools and technologies are appropriate for gathering the most robust data (see Figure 3-1). As described in detail in this chapter, management and interpretation of "big data" will be a continuing challenge for EPA inasmuch as new technologies are now capable of quickly generating huge amounts of data. Senior statisticians are needed in the agency to help analyze, model, and support the synthesis of that data. In many instances, large amounts of data are directly acquired as a component of hypothesis-driven research. However, many new technologies are also used for discovery-driven research— that is, generating large volumes of data that may not be a derivative of a clear, hypothesis-driven experiment, but nevertheless may yield important new hypotheses. In both instances, the data themselves do not become knowledge that can be applied as solutions to problems until they are analyzed and interpreted and then placed in the context of an appropriate problem or scientific theory. As depicted in Figure 3-1, there are iterations and feedback loops that must exist, particularly between data acquisition and data modeling, analysis, and synthesis.

The generation of knowledge, which can take many forms depending on the question being addressed and the nature of the data, ultimately serves as the basis of science-informed regulation and policy (see Figure 3-1). The committee recognizes that scientific data constitute one—albeit important—input into decision-making processes but alone will not resolve highly complex and uncertain environmental and health problems. Ultimately, environmental and health decisions and solutions will also be based on economic, societal, and other considerations apart from science. They need to take into account the variety and complexities of interactions between humans and the environment. But with better scientific understanding, regulations and other actions can be more effective and can have better and more cost-effective outcomes, such as improved human health and improved quality of ecosystems and the environment.

In accordance with the above discussion, it is imperative that EPA have the capacity and knowledge to take advantage of the latest science and technologies, which are always changing. The remainder of the chapter highlights a number of scientific and technologic advances that will be increasingly important for state-of-the-art, science-informed environmental regulation. It also includes several examples of how emerging science, technologies, and tools are transforming the way in which EPA will use data to address important regulatory issues and decision-making, and they demonstrate the need for a systems approach to addressing these complex problems. The chapter has been organized in parallel to the challenges identified in Chapter 2. The main topics that will be discussed are tools and technologies to address challenges related to 1) chemical exposures, human health, and the environment; 2) air pollution and climate

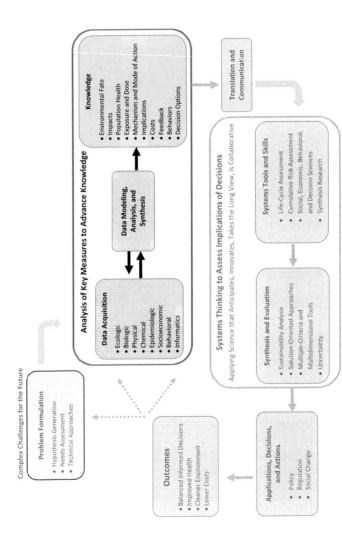

FIGURE 3-1 The iterative process of science-informed environmental decision-making and policy. The process starts with effective problem-formulation, which drives both the experimental design and the selection of data to be acquired. Modeling, synthesis, and analysis of the data are necessary to generate new knowledge. Only through effective translation and communication of new knowledge can science truly inform policies that can generate actions to improve public health and the environment. An evaluation of outcomes is an essential component in determining whether science-informed actions have been beneficial, and it, in turn, adds to the knowledge base.

change; 3) water quality and nutrient pollution; and 4) shifting spatial and temporal scales. The chapter ends with a section on, "Using New Science to Drive Safer Technologies and Products", which discusses ways in which EPA can prevent environmental problems before they arise.

The examples in this chapter are not intended to be comprehensive; rather, they are provided to illustrate from different perspectives the many ways in which new advances in science, engineering, and technology could be embraced by the agency, its scientists, and regulators to ensure that the agency remains at the leading edge of science-informed regulatory policy to protect human health and the environment. Having assessed EPA's current activities, the committee notes that EPA is well equipped to take advantage of most of the new scientific and technologic advances and that, in fact, its scientists and engineers are leaders in some fields.

TOOLS AND TECHNOLOGIES TO ADDRESS CHALLENGES RELATED TO CHEMICAL EXPOSURES, HUMAN HEALTH, AND THE ENVIRONMENT

New technologies will be important to EPA for identifying chemicals in the environment, understanding their transport and fate in the environment, assessing the extent of actual human exposures through biomonitoring, and identifying and predicting the potential toxic effects of chemicals. Current and emerging tools and technologies related to these topics are discussed in the sections below.

Identifying Chemicals in Environmental Media

Analytic chemistry continues to improve at breakneck speed, and analytic determinations for both metals and organic chemicals have improved exponentially. Chemicals can now be detected at ever lower concentrations. For some organic chemicals, such as chlorinated dioxins, standard EPA methods include the routine measurement of samples in parts per quadrillion (ppq) or picograms per liter (pg/L) (EPA 1997), which allows risk managers to characterize lifetime uptake of exposure to various carcinogens and daily uptake rates in chronic hazard quotient assessments of chemicals that were not previously detectable. Simply being able to measure concentrations of chemicals in environmental media or blood confronts EPA with new decisions on whether to set maximum contaminant levels in drinking water or allowable daily intakes in food or whether to allow states to do so independently if health effects are uncertain.

As the public learns about new methods of detection of chemicals in, for example, their blood, their children's blood, and the environment (water, air, and soil), questions arise as to what such occurrences mean. Of course, the simple detection of chemicals in relevant receptors does not necessarily imply any human health or ecologic effects. To evaluate the health implications of chemical exposures throughout the range of exposure levels, sufficiently large epidemiologic studies that incorporate state-of-the-art analytic methods are needed (see the section "Applications of Biomarkers to Human Health Studies"). But, even when biologic effects are not evident (and in special cases of hormesis when there are potentially beneficial effects), the challenge for EPA is to provide meaningful and relevant information to potentially affected parties.

It is now possible, while testing for emerging contaminants of interest and their metabolites, to monitor the effluent of a publicly owned wastewater-treatment plant and determine trace quantities and metabolites of substances—such as pharmaceuticals (licit and illicit), personal-care products, and hormones (natural and synthetic)—that are being used and disposed of or excreted by people in each town (Zegura et al. 2009; Jean et al. 2012; Neng and Nogueira 2012). The mass emission factors per capita can be calculated for the chemicals without determining individual household use. However, without better knowledge of the environmental and human health risks of such low-dose exposures, the advanced detection capabilities do not necessarily help the agency to interpret the results or to protect human health and the environment more effectively. One example is mercury. On one hand, from a toxicologic standpoint, mercury is one of the most studied elements (Schober et al. 2003; Jones et al. 2010). On the other hand, it is still difficult to make a conclusive assessment of the health effects of mercury emitted into the environment (EPA 2011a). Finding cost-effective research opportunities for connecting data on environmental chemicals with environmental and health outcomes can contribute to an increase in knowledge and can inform policy.

Fate and Transport of Chemicals in the Environment

EPA has long been recognized as a leader in developing computer models of the fate and transport of chemical contaminants in the environment, a key component in constructing models of human exposure and health outcomes, as well as in source attribution for ecologic and human endpoints. It develops and supports models for both scientific purposes and application in environmental management. Although many of its models are well established and now backed by years of application experience, EPA and the broader environmental-modeling community face challenges to improve spatial and temporal resolution, to account for stochastic environmental behaviors and for modeling uncertainties, to improve the characterization of transfers between environmental media (air, surface water, groundwater, and soil), and to account for feedback between contaminant concentrations and environmental behavior (for example, the effects of such short-lived radiative-forcing agents as ozone and aerosols have on climate change). Furthermore, sources, properties, and behaviors of some contaminants remain poorly understood, even after years of study. EPA also faces significant challenges and opportunities for integrating models with data from new monitoring systems through data assimilation and inverse modeling techniques. Specific examples of ways in which new approaches to environmental fate and transport modeling are enhancing the understanding of health and ecologic impacts of pollutants are provided in the section on "Tools and Technologies to Address Challenges of Air Pollution and Climate Change".

Assessing the Extent of Human Exposures Through Biomonitoring

Historically, exposure research in EPA has focused on discrete exposures—in external or internal environments, concentrating on effects from sources or effects on biologic systems, and on human or ecologic exposures—one pollutant or stressor at a time. Tools and methods have evolved for undertaking those specific challenges, but targeted approaches have led to sparse exposure data (Egeghy et al. 2012).

The broader availability and ease of use of advanced technologies are resulting in a profusion of data and an overall democratization of the collection and availability of exposure data. The US Centers for Disease Control and Prevention (CDC) National Health and Nutrition Examination Survey (NHANES) alone has provided one of the most revealing snapshots of human exposures to environmental chemicals through the use of biomonitoring (CDC 2012). The collaboration between CDC and national and international organizations quickly expanded the breadth and depth of data available at the population and subpopulation level. That rapid progress was predicated on the availability of better analytic methods and a national commitment to generate baseline data.

Scientific and technologic advances in disparate fields—including computational chemistry, climate change science, health tracking, computational toxicology, and sensor technology—have provided unprecedented opportunities to address the needs of exposure research. Many of the tools are more accessible and easier to use than earlier ones and are slowly being deployed by researchers and stakeholders, such as state agencies and public-interest groups. For example, advances in personal environmental monitoring technologies have been enabled because people around the world routinely carry cellular telephones (Tsow et al. 2009). Those devices may be equipped with motion, audio, visual, and location sensors that can be controlled through wireless networks. Efforts are underway to use them to create expanding networks of sensors to collect personal exposure information.

As discussed in Chapter 2, biomonitoring for human exposure to chemicals in the environment has provided a new lens for understanding population exposures to toxicants. Although the analytic and technical methods discussed to measure human exposure to environmental toxicants will continue to improve, without better information to understand whether the dose is of sufficient magnitude to cause an effect, simply identifying the presence of a toxic substance may raise more questions than it answers. Therefore, there are continuing advances needed to measure and understand the burden of chemicals and their metabolites in the human body.

Recent advances in microchip capillary electrophoresis for separation and identification of nucleotides, proteins, and peptides and advances in spectrometrics, such as nuclear magnetic resonance imaging and mass spectrometry, have changed the nature of health effects monitoring. These technologic advances—especially in genomics, proteomics, metabolomics, bioinformatics, and related fields of the molecular sciences (referred to here collectively as panomics)—have transformed the understanding of biologic processes at the molecular level and should eventually allow detailed characterization of molecular pathways that underlie the biologic responses of humans and other organisms to environmental perturbations. Advances in "–omics" technologies provide EPA with a better understanding of mechanistic pathways and modes of action that can support the risk assessment process. Also, the integration of those technologies with population-based epidemiologic research can contribute to the discovery of major environmental determinants, dose-response relationships, mechanistic pathways, susceptible populations, and gene-environment interactions for health effects in human populations. Appendix C discusses some of the recent advances in -omics technologies and approaches, their implications for EPA, where EPA is at the leading edge of applying the technologies to address environmental problems, and where EPA could benefit from more extensive engagement.

New high-throughput -omic and biomonitoring technologies are providing a greater number of potential biomarkers to assess multiple exposures simultaneously over the course of a lifetime. The biomarkers address exposures to a wide variety of stressors, including chemical, biologic, physical, and psychosocial stressors. The exposome is now being presented as a unifying concept that can capture the totality of environmental exposures (including lifestyle factors, such as diet, stress, drug use, and infection) from the prenatal period on by using a combination of biomarkers, genomic technologies, informatics, and environmental exposures (Figure 3-2) (Wild 2005; Rappaport and Smith 2010; Lioy and Rappaport 2011). The exposome, in concert with the human genome and the epigenome, holds promise for elucidating the etiology of chronic diseases and relevant contributions from the environment (Rappaport and Smith 2010). The concept of the exposome will be of particular value to EPA in assessing and comparing potential health and environmental consequences of individual chemical exposures against previously identified risks. It may also allow for more carefully designed and rational experiments to evaluate potential chemical interactions that contribute to the exposome of individuals or populations.

Exposure information is a key component of prediction, prevention, and reduction of environmental and human health risks. Exposure science at EPA has been limited by the availability of methods, technologies, and resources, but recent advancements provide an unprecedented opportunity to develop higher-throughput, more cost-effective, and more relevant exposure assessments. Research in this field is funded by other federal agencies and international programs, such as the National Institute of Environmental Health Sciences Expo-

Emerging Science & Technologies to Address Environmental Challenges 61

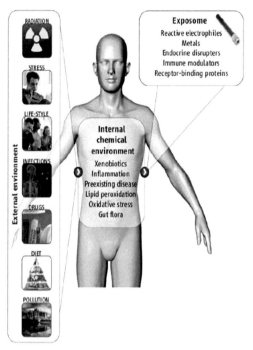

FIGURE 3-2 Characterizing the exposome. The exposome represents the combined exposures from all sources that reach the internal chemical environment. Examples of toxicologically important exposome classes are shown. Biomarkers, such as those measured in blood and urine, can be used to characterize the exposome. Source: Adapted from Rappaport and Smith 2010.

sure Biology Program; the National Science Foundation Environmental, Health, and Safety Risks of Nanomaterials Program; and the European Commission's exposome initiative. Those organizations provide valuable partnership opportunities for EPA to build capacity through strategic collaborations. Moreover, an integral need for EPA in the future will be to develop processes and procedures for effective public communication of the potential public health and environmental risks associated with the increasing number of chemicals, both old and new, that will undoubtedly be identified in food, water, air, and biologic samples, including human tissues. Risk communication strategies should include the latest approaches in social, economic, and behavioral sciences, as discussed in Chapter 5.

Applications of Biomarkers to Human Health Studies

Epidemiologic research plays a central role in assessing, understanding, and controlling the human health effects of environmental exposures. In 2009, the National Research Council (NRC) report *Science and Decisions: Advancing Risk Assessment* recommended that EPA increase the role of epidemiology, surveillance, and biomonitoring to support cumulative risk assessment (NRC 2009). The most successful and current epidemiologic studies leverage multiple resources and use highly collaborative and multidisciplinary approaches (Seminara et al. 2007; Baker and Nieuwenhuijsen 2008). In the United States, a number of high-quality prospective cohort studies funded mostly by the National Institutes of Health have followed millions of people and have collected biospecimen repositories (blood, urine, nails, and DNA) and sociodemographic, genetic, medical, and lifestyle information (Seminara et al. 2007; Willett et al. 2007; NHLBI 2011). Major prospective cohort studies have also been undertaken in other countries (Riboli et al. 2002; Ahsan et al. 2006; Elliott and Peakman 2008).

With some exceptions, current prospective cohort studies generally lack information on environmental exposures. EPA can contribute to closing this gap by, for instance, adding high-quality environmental measures to studies that already have good followup and outcome measures. Examples of collaborations in which EPA plays a critical role are the Agricultural Health Study (NIH 2012), the Multiethnic Study of Atherosclerosis and Air Pollution (MESA Air) (University of Washington 2011), and the National Children's Study (NRC/IOM 2008). In the National Children's Study, the linkage of monitoring data on toxicants in air, water, food, and ecosystems to individual participant data has already been explored in depth in Queens, New York, one of the Vanguard National Children's Study sites (Lioy et al. 2009). Budgetary and implementation challenges for the National Children's Study will require innovative strategies for recruitment, examination, and followup without compromising the quality of the science (Kaiser 2012).

Alternatively, EPA could add followup and outcome measures to studies that have good measures of exposure, although this is likely to be more time-consuming and expensive. At a minimum, EPA should ensure that environmental indicators, including country-wide air-monitoring and water-monitoring data, meet quality and accessibility criteria, for example, through a public data-access system. The indicators can then be merged with individual and community-level data in population-based studies by using geographic and temporal criteria. Biomonitoring and modeling approaches to predict exposure and dose and other advances in exposure science—including the exposome (Weis et al. 2005; Sheldon and Cohen Hubal 2009; Rappaport and Smith 2010; Lioy and Rappaport 2011), -omic technologies, and complex systems approaches (Diez Roux 2011)—could be incorporated into the prospective studies. By building expertise and leadership in exposure assessment and by working in collaboration with other national and international efforts, EPA can play a principal role in the incorporation of environmental exposures into prospective cohort studies and thus contribute to the discovery of major environmental determinants, dose–response relationships, mechanistic pathways, and gene–environment interactions for chronic diseases in human studies.

Environmental informatics plays an important role in the human-population–based studies described above. Although environmental informatics received much of its momentum from central Europe in the early 1990s (Pillmann et al. 2006), EPA has recognized its importance and has played a role in shaping its direction. The agency helped to establish the Environmental Data Standards Council, which was subsumed in 2005 by the Exchange Network Leadership Council (Environmental Information Exchange Network 2011), an environmental-data exchange partnership representing states, tribes, territories, and EPA. The council's mission includes supporting environmental information-sharing among its partners through automation, standardization, and real-time access. The scope of data exchange covers air, water, health, waste, and natural resources, and covers multiple programs. Cross-program data include data from the Department of Homeland Security, the Toxics Release Inventory, pollution-prevention programs, the Substance Registry Services System, and data obtained with geospatial technologies. The council is an example of useful and productive national efforts to generate environmental informatics data. On the basis of technologic advances and new environmental challenges discussed throughout this report, it will be necessary for EPA to begin to make data standards flexible and adaptable so that it can use data that are less structured and less groomed.

Health informatics has a strong history in the United States. There are numerous national and state data registries on chronic and nonchronic diseases, such as the Surveillance, Epidemiology, and End Results cancer registry and the National Birth Defects registry. The Agency for Healthcare Research and Quality of the Department of Health and Human Services maintains a national hospital discharge database and, as previously mentioned, CDC's National Center for Health Statistics conducts the NHANES annually to study health behaviors, dietary intake, environmental exposure, and disease status of the US population. EPA could also work with CDC's National Center for Health Statistics and the National Center for Environmental Health to facilitate the merging of environmental-monitoring data (on air, water, and ecosystems) with national databases that have biomarker and health data, such as NHANES. Such merging, following the NHANES model of public access, could constitute a major advance in the understanding of environmental exposures and their health effects and in informing policy regulation and the prevention and control of environmental exposures. Collaborating with other epidemiologic research efforts, EPA will have the opportunity to identify the optimal population-based prospective cohort study protocol to answer environmental-health questions, to ensure that high-quality data on environmental exposures are incorporated into large epidemiologic studies, and to contribute to the analysis and interpretation of exposure and health-effect associations. In addition, there are proprietary databases owned by healthcare providers and insurers, including Medicare and Medicaid. These databases lay out the foundation of health informatics in the United States and have been successfully used in environmental health research.

Identifying and Predicting the Potential Toxic Effects of Chemicals

In 2007, NRC convened a panel of experts to create a vision and strategy for toxicity testing that would capitalize on the -omics concepts described in Appendix C and on other new tools and technologies for the 21st century (NRC 2007a). Conceptually, that vision is not very different from the now classic four-step approach to risk assessment—hazard identification, exposure assessment, dose–response assessment, and risk characterization—that was laid out in the NRC report *Risk Assessment in the Federal Government: Managing the Process* (commonly referred to as the Red Book) (NRC 1983) and that has been widely adopted by EPA as its chemical risk assessment paradigm (EPA 1984, 2000). However, the vision looks to new tools and technologies that would largely replace in vivo animal testing through extensive use of high-throughput in vitro technologies that use human-derived cells and tissues coupled with computational approaches that allow characterization of systems-based pathways that precede toxic responses. The computational approach to predictive toxicology has many advantages over the current time-consuming, expensive, and somewhat unreliable paradigm of relying on high-dose in vivo animal testing to predict human responses to low-dose exposures.

Although there is generally widespread agreement that the new panomics tools (that is, genomics, proteomics, metabolomics, bioinformatics, and related fields of the molecular sciences), coupled with sophisticated bioinformatics approaches to data management and analyses, will transform the understanding of how toxic chemicals produce their adverse effects, much remains to be learned about the applicability and relevance of in vitro toxicology results to actual human exposures at low doses. With the fundamental mechanistic knowledge, it should be easier to distinguish responses that are relevant to humans from responses that may be species-specific or to identify responses that occur at high doses but not low doses or vice versa. That knowledge would contribute to a reduction in the frequency of false-positive and false-negative results that sometimes plague high-dose in vivo animal testing.

A key issue in the use of such technologies is phenotypic anchoring,[1] which is an important step in the validation of an assay. It is essential to validate treatment-related changes observed in an in vitro –omics experiment as causally associated with adverse outcomes seen in the individual. A single exposure to one dose of one chemical can result in a plethora of molecular responses and hundreds of thousands of data points that reflect the organism's response to that exposure. Quantitative changes in gene expression (transcriptomics), protein content (proteomics), later enzymatic activity, and concentrations of metabolic

---

[1] The concept of phenotypic anchoring arose from studies that examined the effects of chemical exposures on gene expression in tissues (transcriptomics). In that context, the term is defined as "the relation[ship between] specific alterations in gene expression profiles [and] specific adverse effects of environmental stresses defined by conventional parameters of toxicity such as clinical chemistry and histopathology" (Paules 2003).

substrates, products, cofactors, and other small molecules (metabolomics) can all be measured. But which of those signals, if any, are quantitatively predictive of the ultimate adverse response of interest is the key. Changes in the profiles are dynamic, tissue-specific, and dose-dependent, so the results may be drastically different depending on the tissue that was examined, the time when the sample was taken, and the dose or concentration that was used. Sophisticated bioinformatic analyses will be required to make biologic sense out of such massive amounts of data. Tremendous advances have been made in this field in the last 5 years, and it is now possible to coalesce such information into pathway analyses that may have utility in toxicity assessment. Indeed, EPA's ToxCast program has begun to examine approaches discussed above to predictive in vitro toxicity assessment (Judson et al. 2011).

Example of Using Emerging Science to Address Regulatory Issues and Support Decision-Making: ToxCast Program

In 2006, EPA began a new computational toxicology program aimed at developing new approaches to assess and predict toxicity in vitro (Judson et al. 2011). Agency scientists in the computational toxicology program have been substantial contributors to the development of new approaches to toxicity testing. They have collectively published over 130 peer reviewed articles since its inception, including 38 publications from ToxCast (EPA 2012a). Although the use of an array of high-throughput in vitro tests—focused on different putative toxicity endpoints and pathways—to predict in vivo outcomes is attractive from both a cost-savings and time-savings perspective, it entails many challenges, including the following:

- *Chemical metabolism and disposition may differ between the in vitro and in vivo situations.* A principle tenet of toxicology is that the concentration of a toxicant at a specific target site is a key determinant of toxicity. If a metabolite of a toxicant, not the parent molecule, is responsible for toxicity, the in vitro systems must be able to form that metabolite—and other metabolites that might modify the response (for example, alternate detoxification pathways)—in a ratio similar to what occurs in vivo. If an in vitro system fails to form the toxicant or if it forms one that does not occur in vivo, the test system will generate a false-negative or false-positive response. The large amounts of data that can be generated from -omics experiments may be useful in identifying putative pathways of toxicity, but the relevance of the pathways to human exposures depends on a reasonably accurate simulation of the metabolic disposition of the substance that would occur in vivo.
- *The time course of effects observed in vitro may be very different from what occurs in vivo.* Many chemical treatments of cells result in immediate changes in gene expression, and the nature and magnitude of the changes are highly dynamic. Initial responses may be largely adaptive in nature, and not necessarily reflective of an ultimate toxic effect. Adaptive responses can indi-

cate the potential for future toxicity, but many intervening biologic processes may abrogate downstream responses, so the fact that a particular pathway is activated by a chemical does not necessarily mean that the same will occur in vivo. It will be important for high-throughput screening approaches to consider multiple time points for analysis.

- *Dose–response assessment determined in vitro may be difficult to correlate with in vivo responses and administered doses.* Relating dose rate (in milligrams per kilogram per day) in vivo at specific tissues to cell-culture concentrations tested in vitro is extremely difficult and requires detailed knowledge of the absorption, distribution, metabolism, and excretion of a xenobiotic after in vivo exposure. It also requires knowledge about protein binding to plasma and intracellular proteins, lipid portioning, tissue-specific activation, and detoxification for interpretation of the relevance of an in vitro cell concentration to a target-tissue concentration after in vivo administration. Thus, physiologically based pharmacokinetic modeling, which will require some in vivo data, will continue to be an important part of hazard evaluation and risk assessment for chemicals that are identified as being potentially of concern on the basis of in vitro screening assays. Although advances in in vitro toxicity assessment continue to improve and will certainly decrease the number of animals required for in vivo testing, it is unlikely that in vitro tests will fully replace the need for in vivo animal testing for understanding the pharmacokinetics and pharmacodynamics of toxic substances because of the complex interplay between tissues and organs that are ultimately critical determinates of a toxic response.

The importance of those concepts was recently illustrated in some modeling studies of EPA ToxCast data. In the first phase of the ToxCast program, EPA scientists used hundreds of in vitro assays to screen a library of agricultural and industrial chemicals to identify cellular pathways and processes that were modified by specific chemicals; they intended to use the data to set priorities among chemicals for further testing (Judson et al. 2010). However, the potency of a chemical in an in vitro assay may or may not reflect its biologic potency in vivo because of differences in bioavailability, clearance, and exposure (Blaauboer 2010). Scientists at the Hamner Institute, in collaboration with EPA scientists, recently developed pharmacokinetic and pharmacodynamic models that incorporate human dosimetry and exposure data with the ToxCast high-throughput in vitro screening data (Rotroff et al. 2010; Wetmore et al. 2012). Their results demonstrated that incorporation of dosimetry and exposure information is critically important for improving priority-setting for further testing and for evaluating the potential human health effects at relevant exposures.

EPA scientists have played a leading role in the new approaches, and it will be important for them to continue to lead the way in both computational and systems toxicology in the future. With further improvements, such as inclusion of human dosimetric and exposure data, high-throughput in vitro assays for screening of new chemical entities for potentially hazardous properties will probably become widely used for toxicity testing. Although the new technology-driven approaches to in vitro toxicity testing and high-throughput screening constitute an important advance in hazard evaluation of new chemicals, they are not yet ready to replace traditional approaches to hazard evaluation because of inherent limitations of extrapolation from in vitro to in vivo findings, as discussed above. But they will be very useful in setting priorities among new chemicals for more thorough toxicity testing. Additionally, the new technologies will greatly augment traditional approaches to in vivo toxicity evaluation by providing mechanistic insights and more detailed characterizations of biologic responses at doses well below those shown to produce toxicity. That will be especially important in evaluating endocrine-active chemicals and chemically induced alterations that may occur during early life.

McHale et al. (2010) have discussed the importance of new –omics technologies and of a systems-thinking approach to human health risk assessment of chemical exposures, or systems toxicology. EPA has already begun to examine such approaches to predictive in vitro toxicity assessment through the ToxCast program (EPA 2008a). It is evident that new approaches to data management and analysis will be critical for the success of computational approaches to predictive toxicology. The statistical and modeling challenges are immense in addressing the large volumes of data that will come from systems-toxicology experiments, which are an essential element of EPA's computational-toxicology effort. It will be critical for the success of this and other efforts that involve large amounts of data for EPA to have access to the best available tools and technologies in informatics.

Example of Using Emerging Science to Address Regulatory Issues and Support Decision-Making: Predicting the Hazards of a New Material

Nanotechnology is an emerging technology that poses new challenges for EPA. Deemed the next industrial revolution, nanotechnology is predicted to advance technology in nearly every economic sector and be a major contributor to the nation's economy. The rationale of that prediction is that nanoparticles, with dimensions of 1-100 nm, have properties that are useful in a wide variety of applications, including electronic, photovoltaic, structural, catalytic, diagnostic, and therapeutic.

A potential concern is that some of the properties of nanoparticles might pose risks to human health or the environment. The challenge for EPA is to use or develop the science and tools needed to assess and manage the widespread use of nanoscale materials that have unknown hazards. That includes assessing potential risks associated with an emerging technology and, if necessary, monitoring potential exposures and hazards. Using nanotechnology as an example, the committee identified several questions that can be used to better understand the risks associated with new science and tools. Many of these issues regarding the environmental, health, and safety aspects of nanotechnology are addressed in a 2012 NRC report, *A Research Strategy for the Environmental, Health, and Safety Aspects of Nanotechnology* (NRC 2012).

First, do nanoparticles present different properties and inherent risks from smaller molecules or larger particles? To answer this question, it will be important for EPA to adapt and develop new science and tools that strengthen the correlation between the structure and identity of a nanomaterial and the hazard posed by it. That means that new analytic tools or approaches that permit reliable and rapid assessment of engineered-nanomaterial structure and purity are needed. Rapid tests to screen for hazards and set priorities among materials for further testing are essential to keep pace with the development of new materials and to make efficient use of resources available to test materials. To model and predict the properties of the new materials, it will be necessary to develop precisely defined reference materials to ensure that inputs to predictive models and informatics efforts are robust and reliable. The measurement tools, rapid screening approaches, defined reference materials, and modeling and informatics approaches, advanced in an integrated fashion, can determine more rapidly what, if any, unique hazards are associated with this emerging technology.

Second, what are the likely routes and venues of exposure to engineered nanomaterials? Consumer-use patterns, production methods, and life-cycle effects of emerging technologies are unknown. To identify likely ways in which exposure can occur, it will be important for EPA to use physical science, engineering, and social science tools in a multidisciplinary approach that seeks to understand the life cycle of the materials, the supply chains that incorporate them, the projections for market growth, and consumer behaviors in using nanomaterial-containing products. By identifying the intersection between the most likely exposures and unique hazards, EPA can focus on further characterizing the potential risk and using science to inform policies needed to monitor and manage the risk.

Third, how can nanomaterials be detected, tracked, and monitored in complex biologic and environmental media? To complement the science to assess unique hazards and realistic exposures described earlier in this section, EPA will require tools to monitor the distribution of and potential exposures to nanomaterials. The characterization of pristine nanomaterials has been a challenge given the lack of specialized tools for detecting and measuring them. Once distributed, nanomaterials pose even greater challenges to detection, tracking, and monitoring than small molecules or micron-scale particles. This is because nanomaterials tend to have distributions of sizes and surface coatings, their high surface area leads to agglomeration or deposition, their surface chemistry has been shown to be dynamic, and their speciation can be complex. EPA and its collaborators and contractors will need to invent, develop, or refine tools to detect, track, and monitor nanomaterials. In some cases, the solution may be to integrate the use of existing tools. In others, new tools will be required. In addition to direct detection of the materials, strategies that exploit the use of biomarkers as described earlier in this chapter may prove essential for understanding exposures.

The three questions posed in this section may be similarly applied to any emerging material to identify concerns surrounding new hazards, exposure routes, and material tracking. The case of nanotechnology is an example of how EPA will need to approach many emerging tools, technologies, and challenges in general in the future. In order to have the capacity to address those tools, technologies, and challenges, it will need to have enough internal expertise to identify and collaborate with the expertise of all of its stakeholders in order to ask the right questions; determine what existing tools and strategies can be applied to answer those questions; determine the needs for new tools and strategies; develop, apply, and refine the new tools and strategies; and use the science to make recommendations based on hazards, exposures, and monitoring.

TOOLS AND TECHNOLOGIES TO ADDRESS CHALLENGES RELATED TO AIR POLLUTION AND CLIMATE CHANGE

As discussed in Chapter 2, EPA's first goal in its 2011–2015 strategic plan is "taking action on climate change and improving air quality" (EPA 2010). Improved modeling capabilities are integral to attaining that goal inasmuch as models are needed to test the understanding of sources, environmental processes, fate, and effects of airborne contaminants and to investigate the effects of potential mitigation measures. Examples of the many areas in which new technologies will impact air quality and climate change are discussed in the following sections on air-pollution modeling; carbon-cycle modeling, greenhouse-gas emissions, and sinks; and air-quality monitoring.

Air-Pollution Modeling

EPA has a strong history of leadership in air-quality modeling. Its Community Multi-scale Air Quality (CMAQ) model is used both domestically and internationally as a premier platform for "one atmosphere" modeling of the chemistry and transport of ground-level ozone, particulate matter, reactive nitrogen, mercury, and dozens of other materials. In recent years, EPA researchers have worked with other government and university scientists to develop capabilities to run the CMAQ model in a real-time forecast mode (Eder et al. 2009); to couple the CMAQ model to an advanced meteorologic model, the Weather Research and Forecasting system (Appel et al. 2010); and to build advanced sensitivity analysis and inverse modeling capabilities (Napelenok et al. 2008; Tian et al. 2010).

In coming years, investments in modeling efforts will advance the understanding of sources and environmental processes that contribute to particulate-matter loadings and health and environmental effects. Modeling efforts will also improve the understanding of interactions between climate change and air quality with a special focus on relatively short-lived greenhouse agents, such as ozone, black carbon, and other constituents of particulate matter. Improved modeling capabilities will enable EPA to evaluate actions that have dual benefits for reducing radiative-forcing agents (such as ozone and aerosols) and improving air quality, and it will also enable EPA to understand better how tropospheric particulate matter may have masked some global warming in the past. The committee has identified several efforts that will likely be important for EPA in the future. They include, working with other federal and university scientists to improve the use of global climate model predictions to inform air-quality management and other climate-adaptation decisions; working toward a better understanding of the global mass balance of mercury and other biologically active metals, including the role of natural sources and re-emission, chemical and biologic processing, and interregional transport; improving its understanding of physical and chemical processes; leading the integration of models and observations (including satellite and other remote sensing techniques[2]) to help to estimate emissions of greenhouse agents and conventional air pollutants, especially from dispersed or fugitive sources; and expanding its efforts to integrate socioeconomic and biophysical systems models for integrated assessment, including examination of air and climate effects of changing agriculture, energy, information, land-use, and transportation systems.

Carbon-Cycle Modeling, Greenhouse-Gas Emissions, and Sinks

EPA is engaged in a variety of science, engineering, regulatory, and policy development activities related to greenhouse-gas emissions, the global carbon cycle, and impacts of resulting changes on human health. The agency is responsible for the national-level inventory of greenhouse gases in the context of the Framework Convention on Climate Change. Under the Clean Air Act, the agency has authority to regulate greenhouse gases, including carbon dioxide, methane, nitrous oxide, and hydrofluorocarbons. Much attention is also focused on estimating ecosystem uptake and sequestration of carbon as a quantifiable (and monetizable) ecosystem service.

Fossil-fuel emissions can be estimated with relatively high precision, and the science of monitoring and modeling of their uptake by terrestrial and marine ecosystems is evolving rapidly. National-scale and continental-scale estimates of carbon fluxes are now produced through several approaches. In one approach, atmospheric-inversion models rely on regional measurements of atmospheric carbon dioxide coupled to surface ecosystem fluxes and atmospheric circulation

---

[2]Remote sensing—the study of Earth processes and phenomena without direct physical contact—will be discussed several times throughout this chapter. It includes both passive sensors, which measure electromagnetic radiation that is emitted or reflected by the object or area being observed, and active sensors, such as synthetic-aperture radar or light detection and ranging systems, which emit energy and measure its return to infer properties of the scanned surfaces. Remote sensing complements expensive and slow data collection on the ground and provides local-to-global areal coverage of many key environmental processes.

(Gurney et al. 2002). In another approach, which is more direct, biomass inventories (for example, forest and cropland inventories) are used for estimating uptake by monitoring changes in biomass stocks. A third approach involves spatially explicit modeling of ecosystem processes on the basis of weather, soil, land use, and land cover (Schwalm et al. 2010). Each of those approaches has limitations and uncertainties, and derived estimates show only moderate agreement (Hayes et al. 2012). Hayes et al. (2012) demonstrate the value of the inventory approach, which relies on stock estimates obtained from EPA reports (for example, EPA 2011b), for subcontinental-scale estimates of carbon fluxes.

Integrated modeling of greenhouse-gas sources and sinks[3] will continue to develop rapidly given continuing advances in remote sensing of ecosystem properties and understanding of the carbon cycle. To meet its regulatory mandate and to support policies that address climate change, EPA could benefit from increased science and engineering capacity in ecosystem ecology and Earth-system science.

Air-Quality Monitoring

Advances in atmospheric remote sensing have created a new paradigm for air-quality monitoring and prediction from regional to global scales (NRC 2007b). Research and applications have focused on fine particulate aerosols, tropospheric ozone, nitrogen dioxide, formaldehyde, sulfur dioxide, and carbon monoxide, but have also included other compounds, such as benzene, ethylbenzene, and 1,3-butadiene (NRC 2007b, Fishman et al. 2008, Hystad et al. 2011). Active sensors, such as satellite and aircraft-mounted light detection and ranging systems (LiDAR) (for example, the cloud-aerosol LiDAR with orthogonal polarization), can provide information on the vertical distribution of clouds and aerosols on the basis of the magnitude and spectral variation in backscatter of the vertical beam. However, most remote sensing of air quality has relied on passive sensors, for example, measurements of pollution in the troposphere, the moderate-resolution imaging spectroradiometer and multi-angle imaging spectroradiometer on the National Aeronautics and Space Administration's (NASA) Terra platform, and the ozone-monitoring instrument and tropospheric emission spectrometer on NASA's AURA platform (Martin 2008). Those collect radiometric data on solar backscatter or thermal infrared emissions that are then used in retrieval algorithms that incorporate other geophysical information and radiative-transfer models. The reliability of results depends on the surface reflectivity or emissivity, clouds, the viewing geometry, and the retrieval wavelength (Martin 2008). Estimating ground-level concentrations, which are of greatest relevance to EPA, requires additional information on the vertical structure of the

---

[3]The ocean is the largest sink, inasmuch as carbon dioxide is dissolved in seawater and is in equilibrium with the atmosphere (in freshwater bodies, it can change the water pH to some extent).

atmosphere, especially for ozone and carbon monoxide. Inverse modeling is required to infer pollutant source strength from observed concentration patterns.

Although it is not a substitute for ground-based air-quality measurements, satellite-derived data provide important spatial, temporal, and contextual information about the extent, duration, transport paths, and distances of pollution from a source, which is generally not possible with in situ ground-based measurements. For example, Morris et al. (2006) linked increases in surface ozone in Houston to wildfires in Alaska and western Canada, and Heald et al. (2006) traced an increase in springtime surface aerosols in the northwestern United States to anthropogenic sources in Asia. As retrieval algorithms and the spatial and spectral quality of satellite data have improved, remote sensing has provided a means of obtaining relatively consistent estimates of air-pollutant exposure over large areas for health-effects assessments (van Donkelaar et al. 2010; Hystad et al. 2011), which has facilitated large-scale epidemiologic investigations in settings where monitoring data are inadequate to determine spatial contrasts (Crouse et al. 2012). Another important trend is the assimilation of concurrent data from multiple sensors with ground data; that has proved especially useful in improving estimates of ground-level ozone (Fishman et al. 2008).

Example of Using Emerging Science to Address Regulatory Issues and Support Decision-Making: Remote Sensing to Monitor Landfill Gas Emissions

Great progress has been made in reducing or eliminating releases of toxic substances from concentrated sources (also known as point sources), but monitoring and mitigating emissions from so-called area sources has been technically difficult and remains one of the persistent challenges faced by EPA. Recent efforts to use emerging technology in monitoring provide a glimpse on a very broad scale of what might be possible with further advances. EPA's National Risk Management Research Laboratory used a tunable diode laser to perform optical remote-sensing of fugitive methane, hazardous air pollutants (including mercury), volatile organic compounds, and nonmethane organic compounds emitted from three landfills. With multiple measurements of concentrations along different light paths, the system calculates a mass emission flux for the entire area. What had been thought to be an excessively expensive monitoring challenge is now financially and practically manageable (EPA 2012b).

Example of Using Emerging Science to Address Regulatory Issues and Support Decision-Making: Multipollutant Analysis Standard-Setting

Regulation in the United States is predicated on single-pollutant standards or control strategies. Improved understanding of health effects of cumulative and mixed exposures calls for new approaches to standard-setting that consider a multipollutant approach. The shift will require understanding of the joint behavior of multiple stressors, the interactions among them, and their contributions to health outcomes.

The air-pollution health community has been examining the science-readiness of a multipollutant regulatory strategy (Dominici et. al. 2010; Greenbaum and Shaikh 2010). The challenges, opportunities, and future research needs related to multipollutant approaches for the assessment of health risks associated with exposures to air pollution were evaluated in a public workshop held in 2011 (Johns et al. 2012). The workshop highlighted the need for a transdisciplinary research approach for developing more relevant tools and methods in the fields of exposure science, human and animal toxicology, and air pollution epidemiology. More important, it recommended collaboration among science, engineering, and policy communities to develop practical and implementable approaches that could ultimately inform decision-making (D. Johns, EPA, personal communication, May 9, 2012).

Related efforts to characterize toxicity of mixtures of chemicals in chemical risk assessment are under way. A key challenge is to define the universe of possible combinations of mixtures that are representative of real-world exposures. In a recent analysis, EPA researchers investigated methods from the field of community ecology originally developed to study avian species co-occurrence patterns and adapted them to examine chemical co-occurrence (Tornero-Velez et al. 2012). Their findings showed that chemical co-occurrence was not random but was highly structured and usually resulted in specific predictable combinations. Novel application of tools and approaches from a variety of research disciplines can be used to address the complexity of mixtures, advance the scientific communities' understanding of exposures to the mixtures, and promote the design of relevant experiments and models to assess associated health risks.

TOOLS AND TECHNOLOGIES TO ADDRESS CHALLENGES RELATED TO WATER QUALITY

As discussed in Chapter 2, there are several important drivers of water quality and water-quality policies for which new technologies and approaches can be instrumental in enhancing data-driven regulations. For the purposes of this chapter, examples of the many areas in which new technologies will impact water quality are divided into the following areas: remote sensing technologies for water-quality monitoring; water modeling; and detecting microorganism and microbial products in the environment.

Water-Quality Monitoring

Multispectral imagery has been successfully applied to water-quality monitoring for several decades, notably for monitoring surface temperature and concentrations of suspended sediments and algae (see reviews by Mertes 2002; Matthews 2011). Modern multispectral sensors—such as the moderate-resolution imaging spectroradiometer and the European medium-resolution imaging spectrometer sensor, which have moderate (about 250m) spatial resolution, 10–15 spectral bands, and high sampling frequency—have accelerated progress in remote sensing of suspended sediments, dissolved organic matter, chlorophyll, phycocyanin, and other water-quality indicators that are extensive enough to suit sensor resolution (Bierman et al. 2011). Satellite-based assessments of water quality will probably be increasingly routine, especially with better integration and assimilation of in situ data and multiscale sensor data via empiric and physically based models (Matthews 2011). As mentioned in the section "Air-Quality Monitoring" above, the new tools and technologies are not a substitute for ground-based water-quality measurements, but they provide important spatial and contextual information about the extent, duration, transport paths, and distances of pollution from a source and should be used to enhance the current water-monitoring infrastructure and related exposure assessment efforts.

Water Modeling

Real-time reporting of water-quality data would complement EPA research programs. Data could be downloaded to a community Website so that other researchers and the general public could understand water-quality and quantity (storm-flow) information better. That type of network would eventually allow analysis of infiltration or inflow problems, including policy options (such as disconnecting storm drains from the sanitary sewer) and the likely effectiveness of infrastructure investment in light of climate change (such as more intense storm events). Figure 3-3 illustrates how a sensor network might be set up.

Spatially detailed high-frequency sensing of water resources that uses an embedded network can provide breakthroughs in water science and engineering by promoting understanding of nonlinearities (the knowledge base to discern mechanisms and basic kinetics of nonlinear water processes) (Ostby 1999; Coppus and Imeson 2002; Nowak et al. 2006); scalability (the ability to scale up complex processes from observations at a point to the catchment basin) (Ridolfi et al. 2003; Sivapalan et al. 2003; Long and Plummer 2004); prediction and forecasting (the capacity to predict events, to model and anticipate outcomes of management actions, and to provide warnings or operational control of adverse water-quantity and water-quality trends or events) (Christensen et al. 2002; Scavia et al. 2003; ASCE 2004; Vandenberghe et al. 2005; Shukla et al. 2006; Hall et al. 2007); and discovery science (the discovery of heretofore unknown and unreported processes) (Jeong et al. 2006; Messner et al. 2006; Loperfido et al. 2009; 2010a,b).

Detecting Microorganisms and Microbial Products in the Environment

Development of detection methods for microbial contamination in water, soil, and air is a critical part of environmental protection. EPA is one of the few federal agencies that oversees a substantial research portfolio that includes new analytic techniques for environmental assessment. Although more modern biochemical methods are available, coliform bacteria and enterococci continue to be used as indicators for the assessment of safe drinking and recreational waters (EPA 1986, 2002, 2005), and cultivation methods for viability remain the gold standard (Messer and Dufour 1998). In recognition of the inadequacy of the bacterial indicator system over the years, research methods have been developed and improved for measuring enteric viruses (Fong and Lipp 2005; Yates et al. 2006; Pepper et al. 2010) and protozoa (Sauch et al. 1985; Rose 1988; Aboytes et al. 2004) and for expanding the understanding of risk (Slifko et al. 1997, 1999; Aboytes et al. 2004). National surveys of groundwater and surface water have directly influenced important rule-making, including the Surface Water Treatment Rule, the Information Collection Rule, the Long Term Enhanced Surface Water Treatment Rule, the Disinfectants and Disinfection Byproducts Rule, the Ground Water Rule, and final rules for the use or disposal of sewage sludge.

Assessment and control of waterborne diseases still rely on the ability to sample and quantify fecal indicator organisms and pathogens as part of the evaluation of water quality. The most recent advances in the detection of microorganisms in water include quantitative polymerase chain reaction (PCR) methods, which can be designed for any microorganism of interest because they are highly specific and quantitative. The PCR methods can produce information relatively fast and, under the Clean Water Act and the Beaches Environmental Assessment and Coastal Health Act, their adoption has moved quickly toward meeting total maximum daily load requirements and beach safety (see the example below on "Beach Safety").

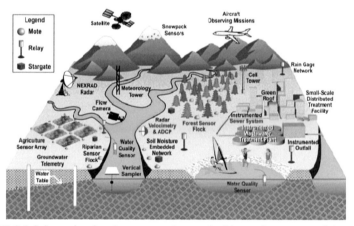

FIGURE 3-3 Schematic of an instrumented watershed in an observatory of the national network. Real-time sensors for meteorology, rainfall, stream velocity, suspended sediment, water quality, soil moisture, groundwater, and snowpack are shown with wireless communications equipment necessary for transmitting the data. Source: WATERS Network 2009.

New approaches to next-generation DNA-sequencing technologies offer the promise of characterizing healthy water by ensuring the absence of harmful biotic organisms, even rare ones (see Appendix C for background information on genomics tools and technologies). Just as the human microbiome studies are examining the diversity and ecology of microorganisms in the intestinal tract, DNA-sequencing methods are being used to explore the water microbiome in polluted, pristine, and unique environments, although finding rare microbial populations that will exhibit genetic characteristics with the potential for harm to humans is difficult. Metagenomics of the wastewater system, and in particular the viral genome, provide insight into the complex world of water microbiotas, but is only being used for exploration. Current efforts are being spent in developing methods and generating large amounts of data (Table 3-2); the methods are able to identify which microorganisms (including potentially pathogenic organisms) are present, but their viability and functional activity are often not known. Finally, genomic data have not been used much to inform microbial risk assessment. In the next decade, environmental microbiome studies and data will need to move toward sophisticated data interpretation and modeling, and substantial investment in bioinformatics will be necessary. With the growing understanding of the ecosystem microbiome and its interaction with human health and the environment, it is becoming evident that the microbiome plays an important role in modulating health risks posed by broader environmental exposures. Understanding such interactions will have important implications for understanding individual and population susceptibility and the observed variability in risks posed by environmental exposures.

Other recent advances that are facilitating the use of molecular tools include new techniques for increasing sample concentration—such as ultrafiltration, continuous filtration, and new types of filters—for improved recovery and automated extraction of nucleic acids with less contamination, less inhibition, and more rapid throughput (Hill et al. 2005; Srinivasan et al. 2011). New quantitative PCR approaches for monitoring the viability of pathogens of concern are of particular interest, and several approaches show some promise. Such dyes as ethidium monoazide and propidium monoazide have been used to distinguish between live cells and heat-killed cells, but the dyes are not able to penetrate apparently killed cells when applied to disinfected treated sewage samples, so the signals that are produced through quantitative PCR methods are comparable with counts made before and after disinfection with or without use of the dyes (Varma et al. 2009; Srinivasan et al. 2011). More work is needed to address the possible presence of viable but nonculturable cells in disinfected effluents. An approach to examining viability associated with bacteria is to use quantitative PCR methods to target the precursors of ribosomal RNA (rRNA). That was done to quantify viable cells of *Aeromonas* and mycobacteria in water (Cangelosi et al. 2010) and showed promise for both saltwater and freshwater and for post-chlorination monitoring. Those types of methods will require verification in the monitoring of disinfected drinking water and wastewater. There may be a need

TABLE 3-2 Metagenomic Characterization of Pathogens and Microbial Populations in Biosolids, Wastewater, Rivers, and Lakes

Environment Sampled	Target and Approach	Findings	Reference
Wastewater biosolids	Bacterial 16S rRNA genes; PCR, pyrosequencing (454 GS-FLX sequencer)	Most of the pathogenic sequences belonged to the genera *Clostridium* and *Mycobacterium*	Bibby et al. 2010
Wastewater (activated sludge, influent, and effluent)	Bacterial 16S rRNA gene (hypervariable V4 region); PCR, pyrosequencing (454 GS-FLX sequencer)	Most of the pathogenic sequences belonged to the genera *Aeromonas* and *Clostridium*	Ye and Zhang 2011
River sediment	Bacterial antibiotic-resistance genes; MDA, pyrosequencing (454 GS-FLX sequencer)	Large amounts of several classes of resistance genes in bacterial communities exposed to antibiotic were identified	Kristiansson et al. 2011
Reclaimed and potable water	Viral DNA and RNA; tangential flow filtration, DNase treatment, MDA, pyrosequencing (454 GS-FLX and GS20 sequencer)	Over 50% of the viral sequences had no significant similarity to proteins in GenBank; bacteriophages dominated the DNA viral community; the RNA metagenomes contained sequences related to plant viruses and invertebrate picornaviruses	Rosario et al. 2009
Wastewater biosolids	Viral DNA and RNA; DNase and RNase treatment, reverse transcription for RNA, pyrosequencing (454 GS-FLX sequencer), optimal annotation approach specific for viral pathogen identification is described	Parechovirus, coronavirus, adenovirus, aichi virus, and herpesvirus were identified	Bibby et al. 2011
Lake water	Viral RNA; tangential flow filtration, DNase and RNase treatment, random amplification (klenow DNA polymerase), pyrosequencing (454GS-FLX sequencer)	66% of the sequences had no significant similarity to known sequences; presence of viral sequences (30 viral families) with significant homology to insect, human, and plant pathogens	Djikeng et al. 2009

Abbreviations: DNase, deoxyribonuclease; MDA, multiple displacement amplification; PCR, polymerase chain reaction; RNase, ribonuclease; rRNA, ribosomal RNA.

Source: Aw and Rose 2012. Reprinted with permission; copyright 2012. *Current Opinion in Biotechnology*.

for a method that combines some type of cultivation with quantitative PCR techniques in real time to address viability. The use of molecular tools that can be used to inform decisions for water treatment and public-health protection will still require substantial investment in sample concentration, hazard characterization, quantification, and assessment of viability.

Example of Using Emerging Science to Address Regulatory Issues and Support Decision-Making: Beach Safety

Shorelines provide benefits to society as a whole and in particular are directly associated with tourism, which remains one of the largest economic sectors around the world. According to the Natural Resources Defense Council (NRDC 2011), beaches in the United States were given advisories or were closed 24,091 times in 2010—the second-highest number of advisories and closures in the 21 years since NRDC began reporting. It was suggested that aging and poorly designed sewage-treatment systems and contaminated stormwater were the main causes of pollution that led to fecal–indicator concentrations that exceeded the state's health and safety standards. There were also more than 9,000 days of Gulf Coast beach notices, advisories, and closures due to the Deepwater Horizon oil-spill disaster in 2010 (NRDC 2011).

As part of an overhaul of the Clean Water Act, the Beaches Environmental Assessment and Coastal Health Act mandated that research be undertaken to understand coastal pollution, address polluted sediments, decrease response time, and improve protection of public health (EPA 2006a); most of the research programs under this act have yet to be realized, and improving public-health protection has been slow. EPA has begun to update water-quality standards, address health studies and swimmer surveys, and advance the use of new genomic technology for the rapid testing of water quality. The development of the first standardized quantitative PCR method for enterococci is being promoted for recreational-water assessment (Wade et al. 2006). Evaluations based on new quantitative PCR methods for indicators in ambient and recreational waters are being published (Byappanahalli et al. 2010; Noble et al. 2010), but there are challenges to using these methods for regulatory purposes because interpretation of the signals may underestimate or overestimate human health risks and could lead to beach closures that cause unnecessary economic losses (Srinivasan et al. 2011). Continued investment in new methods, applications for surveys, and links to health effects and management strategies are necessary.

Wastewater and stormwater are key culprits in water pollution, and further improvement of water safety cannot occur unless point and nonpoint sources of pollution are elucidated. Research on microbial source tracking has advanced the use of molecular tools for investigating the presence of pathogens in impaired waters and to setting total maximum daily load requirements. EPA is taking a leadership role in the microbial source-tracking research (EPA 2005). In addition, California has organized one of the largest blind studies, the Global

Inter-Laboratory Fecal Source Identification Comparison Study, which involves the evaluation of 39 microbial source-tracking methods by 29 laboratories (Shanks 2011).

To maximize the benefits of clean water, protect the general public, sustain water resources, and restore impaired shorelines, decision-makers will need to rely increasingly on an understanding of the long-term and short-term changes in water quality and aquatic ecosystems. The advanced science and technology are poised to play an increasingly important role in providing forecasts of effects on appropriate temporal and spatial scales. Advances could be made quickly for safe and sustainable water resources in the promotion of methodologic developments and applications in rapid and predictive monitoring; development of and investment in a safe-waters program that links genomic tools with watershed and beach-shed characterizations; continued microbial characterization of stormwater, combined sewage overflows, and wastewater; and development of and investment in innovative engineering designs to reduce pollution loads.

Example of Using Emerging Science to Address Regulatory Issues and Support Decision-Making: Quantitative Microbial Risk Assessment

Quantitative microbial risk assessment had its beginnings in the 1980s; it is associated with the first publication of dose–response models (Haas 1983) and is now an accepted process for addressing waterborne disease risks and management strategies (Haas et al. 1999; Medema et al. 2003). Although great strides have been made in using quantitative microbial risk assessment in EPA's Office of Homeland Security (including leading an interagency working group and the exchange of information with CDC), EPA has yet to take a leadership role in developing the necessary databases for use in a national risk assessment of wastewater, stormwater, and recreational water.

Linking biology, mathematics, health, the environment, and policy will require substantial interdisciplinary research focused on problem-solving and systems thinking. Quantitative microbial risk assessment has been seen as an important framework for pulling science and data together and can lead to innovative work in decision science. According to the Center for Advancing Microbial Risk Assessment, "ultimately, the goal in assessing risks is to develop and implement strategies that can monitor and control the risks (or safety) and allows one to respond to emerging diseases, outbreaks and emergencies that impact the safety of water, food, air, fomites, and in general our outdoor and indoor environments" (CAMRA 2012). The framework is being promoted by the World Health Organization (WHO 2004), and the international need for data, education, and mathematical tools to assist countries around the world with the implementation of quantitative microbial risk assessment strategies is paramount. More recently, *Science and Decisions: Advancing Risk Assessment* (NRC 2009) called for more integration with the risk-assessment–risk-management paradigm. This approach will provide a pathway to the integration of new tools and science for addressing EPA's goals of safe and sustainable water.

If a quantitative microbial risk-assessment framework were put into practice by EPA, it would need to incorporate alternative indicators based on genomic approaches, microbial source-tracking, and pathogen-monitoring. Also, the complete human-coupled water cycle would need to be explored, including built and natural systems. Implementation of a quantitative microbial risk-assessment framework would require investment in a health-related water microbiology collaborative research network. The network would bring molecular biologists, ecologists, engineers, and water-quality health and policy experts together to build internal capacity, to develop external partnerships, and to foster national collaboration. Regardless of whether EPA decides to systematically use a quantitative microbial risk-assessment framework, the future of science at the agency would benefit from continuing to build exposure databases and support work on the survival and inactivation of pathogens that can feed into quantitative microbial risk assessment. Agency science would also benefit from new informatics and application tools that are based on quantitative microbial risk assessment models to enhance decision-making to meet safe-water goals.

An example of an area in which EPA may be able collaborate to more effectively fill information gaps or address funding overlap in a resource-constrained environment is through microbiology research. There are other organizations that have microbiology programs, but few address the environment. NIH's Division of Microbiology and Infectious Diseases supports clinical research and basic science for microbes and infectious disease. NIH has recently partnered with the National Science Foundation and the US Department of Agriculture to fund research on the ecology and evolution of infectious disease. The partnership addresses diseases that have an environmental pathway and can include waterborne diseases, but most of the efforts have been related to cholera and little attention has been given to other groups of pathogens. EPA has not yet played a role in the partnership, but it could contribute to filling a gap in knowledge about wastewater treatment and monitoring as it relates to microbes and environmental and human health.

TOOLS AND TECHNOLOGIES TO ADDRESS CHALLENGES RELATED TO SHIFTING SPATIAL AND TEMPORAL SCALES

Chapter 2 noted that current environmental challenges are expanding in both space and time and it emphasized that long-term data are needed to characterize such changes and to characterize the cause and the potential implications of different policy options. To address the challenges of increasing spatial and temporal scales for a variety of environmental problems, new approaches, tools, and technologies in such areas as computer science, information technology (IT), and remote sensing will become increasingly important to EPA. The ability to take full advantage of all the new tools and technologies discussed in the preceding sections of this chapter will require EPA to have state-of-the-art IT and informatics resources that can be used to manage, analyze, and model diverse datasets obtained from the vast array of technologies.

Computer Science, Informatics, and Information Technology

The future needs for IT and informatics in support of science in EPA are subject to two principal influences: the future directions of EPA's mission and the underlying science in future directions taken by the IT industry. Science in EPA will increasingly depend on its capability in IT and informatics. IT is concerned with the acquisition, processing, storage, and dissemination of information with a combination of computing and telecommunication (Longley and Shain 1985). The term *informatics*, as used here, refers to the application of IT in the generation, repository, retrieval, processing, integration, analysis, and interpretation of data obtained in different media and across geographic and disciplinary boundaries that are related to the environment and ecosystem, community and human activities, and human health (see He 2003). Informatics is also concerned with the computational, cognitive, and social aspects of IT. One way in which IT can be used for data acquisition is through public engagement. Taking advantage of expertise outside of EPA (from academia, industry, and other agencies) and considering the general public as a source of new information is a way in which knowledge and resources can be combined in a cost-effective manner. Examples include taking advantage of social media and crowdsourcing. Appendix D provides additional background information on various important and rapidly changing tools and technologies in the field of information technology and informatics.

Example of Using Emerging Science to Address Regulatory Issues and Support Decision-Making: Social Media

EPA does substantial outreach to the public and to other agencies and research communities via such media as blogs and wikis. It also supports mobile, desktop, and laptop collaboration and it clearly sees the role of social media for these outward-facing purposes. The general IT activities are the responsibility of several entities in the Office of Environmental Information and elsewhere in the agency, such as the Office of Solid Waste and Emergency Response (EPA 2012c) and the Office of Water (EPA 2012d). Social media also have a role to play in crowdsourcing and citizen science, as will be discussed in the following section. Another important topic in the near future will be the use of social media for scientific collaboration. The emergence of secure enterprise social networks provides a host of opportunities for greatly enhanced internal and external collaboration, particularly as tighter budgetary circumstances force the dissolution of some departmental and interagency boundaries.

Those networks already securely provide an environment for microblogging, private messaging, profiles, administered groups, directories, and secure external networks of partners. Conversations may be fully archived and are searchable with tags, topics, and links to documents and images. The technology is accelerating rapidly and will surely be part of the expectations for the next generation of EPA scientists. As the new technologies are emerging, consolidating, and maturing, following such changes closely would help EPA to make anticipatory decisions for adopting the appropriate technology that provides the greatest benefit to the agency at the least cost.

Example of Using Emerging Science to Address Regulatory Issues and Support Decision-Making: Crowdsourcing

Massive online collaboration, or crowdsourcing, can be defined as the "sourcing [of] tasks traditionally performed by specific individuals to an undefined large group of people [or community]—the crowd—through an open call. For example, the public can be invited to help develop a new technology, carry out a design task, [propose policy solutions,] or help capture, systematize, or analyze large amounts of data—also known as citizen science" (Ferebee 2011). With a well-designed process, crowdsourcing can help assemble the data, expertise, and resources required to perform a task or solve a problem by allowing people and organizations to collaborate freely and openly across disciplinary and geographic boundaries. The emergence of crowdsourcing, such as citizen science, with widely dispersed sensors will produce vast amounts of new data from low-cost unstructured sources. This can inform multiple domains of environmental science, but may have the greatest potential for monitoring environmental conditions and creating more refined models of human exposure.

The idea behind regulatory crowdsourcing is that many areas of regulation today, from air and water quality to food safety and financial services, could benefit by having a larger number of informed people helping to gather, classify, and analyze shared pools of publicly accessible data. Such data can be used to educate the public, enhance science, inform public policy-making, or even spur regulatory enforcement actions. Indeed, there are many arenas in which experts and enthusiasts, if asked, would help to provide data or to analyze existing data.

EPA is no stranger to crowdsourcing. With such peers as NASA and CDC, EPA is a pioneer and visible leader in collaborative science. One example is its use of broad networks to engage outside environmental problem-solvers, for example, through the Federal Environmental Research Network, InnoCentive challenges, and the Challenge.gov Web site (Preuss 2011). More recent examples of using the public to gather information are discussed in Box 3-1. Broad community participation has led to a wide array of new data sources to provide a baseline for monitoring the effects of climate change on local tree species, for wildlife toxicology mapping, and for real-time water-quality monitoring.

Today, a growing number of regulatory agencies (including EPA, the Securities and Exchange Commission, and the Food and Drug Administration) see social media and online collaboration as a means of providing richer, more useful, and more interactive pathways for community participation. For EPA and its stakeholders, the question is whether the agency can take advantage of this growing social interconnectivity to engage the public in environmental protection better while bolstering both its science activities and its capacity for effective regulatory monitoring and enforcement. There are a number of ways in which crowdsourcing or citizen science could augment or enhance EPA's scientific and regulatory capabilities. They include harnessing new technologies to engage broader communities along the lines of crowdsourced data collection, especially in the context of environmental monitoring, exposure assessment, health surveillance, and social behaviors; crowdsourced data classification and analysis; and crowdsourced environmental problem-solving. Crowdsourcing also provides an opportunity for EPA to gain a better understanding of the general sentiment of the public on issues that are of concern to EPA.

Crowdsourcing initiatives are typically low in cost because the most expensive resource (people's time) is supplied voluntarily. Whether classifying galaxies or recording observations of bird species or local environmental quality, participants in a crowdsourcing project are intrinsically motivated to participate. For an agency like EPA, crowdsourcing presents an opportunity to gather and analyze large amounts of data or input inexpensively. That being said, crowdsourcing projects are not free to run either. There are costs involved in supplying the infrastructure for participation (typically a Web site or mobile interface where participants can record observations and discuss issues) and managing the overall effort.

Acquisition of Environmental Data through Remote Sensing

In the 40 years since the launch of Landsat 1—the first Earth-observing satellite-borne sensor designed expressly to study the planet's land surfaces—there have been enormous advances in remote-sensing systems for environmental mapping and monitoring. They include multispectral digital imaging systems and imaging radar (1970s), hyperspectral imaging systems (1980s), and profiling and imaging LiDAR (1990s to present). In that period, remote sensing has benefited from rapid improvements in instrument capabilities and calibration, positional control and global positioning systems, computer performance, processing algorithms and software, fusion of imagery from multiple sensors, and closer integration with geographic information system and ground measurements and monitoring systems. As a result, remote sensing of the environment has evolved from a narrow research community to a large and diverse user community that is applying remote-sensing products on local, regional, and global scales (Schaepman et al. 2009).

> **BOX 3-1** Engaging the Public to Gather Information
>
> One example of crowdsourcing is a "computer game" called Fold-IT, in which participants work to fold proteins in different configurations (Foldit 2012). The scoring for the game is based on packing the protein (the smaller the better), hiding the hydrophobic side chains, and folding the protein so that sidechains are not too close together. The information gathered from this program has resulted in the publication of several scientific papers (Cooper et al. 2010a,b, 2011; Gilski et al. 2011; Khatib et al. 2011a,b) and has shown that sometimes human knowledge and intuition can outperform computational methods.
>
> Another example is when EPA set out to produce an action plan for the Puget Sound estuary in Washington state. It launched an information challenge that invited the broader community to assemble relevant data sources and begin to articulate solutions. Over 600 residents, businesses, environmental groups, and researchers contributed 175 new data sources (Tapscot and Williams 2010). Examples include a tree-ring database from 2006 that provides a baseline for monitoring the effects of climate change on local tree species, wildlife toxicology maps of the Puget Sound area, and real-time water quality-monitoring tools, including water measurements taken from local ferries that could complement existing buoy measurement systems. Former EPA Chief Information Officer Molly O'Neill said afterwards, "we can actually use these kinds of mass collaboration tools to transform government, not just add layers to government" (Tapscot and Williams 2010). The kinds of "emergent behavior" observed in cases like the Puget Sound information challenge can be applied in nearly all aspects of the regulatory system and lead to new insights, innovations, and strategies that even the most capable agencies could not produce in isolation (PSP 2011).

EPA has long recognized the scientific value and cost effectiveness of remote sensing for large-area environmental mapping and monitoring, and remote sensing data are increasingly being used to strengthen human exposure characterization for air pollutants and other contaminants. The agency has been a contributing partner in national satellite-based mapping programs such as the Multi-Resolution Land Cover Program, the Coastal Change Analysis Program, and the Gap Analysis Program. It has also supported research and application efforts in remote sensing of water quality and air quality, notably in the use of aircraft-borne sensors for local pollution and hazardous-substance detection and monitoring. For example, from 2002 to 2010, the agency partnered with the Department of Defense and the National Geospatial Intelligence Agency to operate Airborne Spectral Photometric Environmental Collection Technology, an aircraft-borne set of sensors designed to provide emergency-response data on hazardous releases.

Much of the progress in remote sensing has depended on tight integration of imagery with other geospatial data and process models that use appropriate parameters and aircraft and satellite data. A key opportunity for EPA science lies in extended collaboration with remote-sensing scientists to advance such integrated approaches, especially in fields in which EPA has extensive or nascent expertise in such domains as pollutant fate and transport, landscape ecology, ecosystem service mapping and monitoring, environmental-disaster monitoring, and health impact assessment.

Terrestrial-Ecosystem Monitoring with Remote Sensing

Remote sensing of land surfaces has evolved from technically complex but thematically relatively simple land-use and land-cover mapping and monitoring to technically and scientifically complex monitoring and modeling of surface properties and processes, such as three-dimensional (3D) vegetation structure and net primary production. From a technologic perspective, important trends in remote sensing of terrestrial ecosystems include (Wang et al. 2010)

- Increasing availability of multispectral imagery with very high spatial resolution (0.5–10 m) from satellite systems such as IKONOS, GeoEye-1, SPOT-5, and FORMOSAT-2.
- Increasing availability of imaging spectrometer data with more than 100 narrow (10–20 nm) spectral bands at moderately high (10–500 m) resolution from satellite-borne systems, such as EO-1 Hyperion, and aircraft-borne systems, such as the Compact Airborne Spectrographic Imager.
- Imaging LiDAR from aircraft platforms for regional studies (for example, laser vegetation-imaging sensors) and satellite platforms for global studies (for example, the Geosciences Laser Altimeter System carried on the Ice, Cloud, and Land Elevation Satellite).
- Well-calibrated thermal remote-sensing data at fine spatial resolution (the Advanced Spaceborne Thermal Emission and Reflection Radiometer and the Hyperspectral Infrared Imager), moderate resolution (the Moderate Resolution Imaging Spectrodiameter and the Visible Infrared Imager Radiometer Suite), and coarse resolution (the Geostationary Operational Environmental Satellite and the Meteosat Second Generation Satellite) for monitoring surface-energy balance, evapotranspiration, plant stress, and drought.
- Constellations of small satellites capable of high-frequency global coverage for environmental event and disaster monitoring (for example, the UK Disaster Monitoring Constellation Satellite).

Imaging spectrometers hold special promise for obtaining detailed information about plant-community composition and the physiologic condition of canopies and allowing monitoring of community succession, phenology, species invasions, crop yield, soil chemistry, and nutrient cycling. Issues of data quality and data access are diminishing, and progress is being made in radiative-transfer models, spectral-mixture models, and physically based inversion models for multiscale monitoring of terrestrial ecosystem processes (Schaepman et al. 2009).

Imaging LiDAR is especially powerful for tracking changes in surface elevation, above-ground vegetation biomass, 3D vegetation structure, and 3D distribution of canopy-leaf area. Particularly in forested regions, LiDAR can be used to improve estimates of net ecosystem production and carbon stocks over large areas (Goetz and Dubayah 2011; Hall et al. 2011). Imaging radar has also been an important tool for monitoring vegetation structure, especially in cloud-prone areas and areas subjected to seasonal flooding (Bergen et al. 2009). Image fusion, the combined use of imagery from two or more sensors, can be used to exploit complementary information from very-high-resolution multispectral imagery, hyperspectral imagery, and LiDAR (Koetz et al. 2007).

Long-Term Datasets in Real-Time

Dense, long-term environmental datasets in real-time could create a foundation for informed decision-making. A suite of decision-support tools could be developed for integration with air- and water-quality models at various scales. For example, data in hospital admission forms could be combined with meteorological and air quality models in real time to provide health forecasts and warnings. Real-time sensing and modeling of water-borne pathogens in situ could provide drinking water treatment plants with threat forecasts, alerting them to the need to change source water or treatment techniques. Special research attention could be given to handling uncertainty of both data and models. In addition to deterministic models with uncertainty analyses, probabilistic approaches can be extremely powerful when computational intelligence tools are used.

Data assimilation and data mining approaches provides innovative possibilities. An example is the use of an intelligent real-time cyberinfrastructure-based information system called the Intelligent Digital Watershed to better understand the interactions and dynamics between human activity and water quality and quantity. Such an approach provides "1) novel uses of data mining algorithms in data quality and model construction, 2) development of specialized data mining algorithms for [environmental forecasting] applications, 3) development of data transformation algorithms, [4)] data-driven modeling of non-stationary processes, [such as storm forecasting for by-pass wastewater discharges], and [5)] development of decision-making algorithms for models constructed with data mining algorithms".[4] Using data in a novel way could greatly expand the analysis capability of EPA and provide insights previously impossible to obtain without such innovations.

Already, the Consortium of Universities for the Advancement of Hydrologic Science, Inc. Hydrologic Information System (HIS) project has a systematic data acquisition network for the publication, discovery, and access of water

---

[4]NSF-CDI. 2008-2011. CDI-Type II: Understanding Water-Human Dynamics with Intelligent Digital Watersheds. (#0835607). Jerald L. Schnoor (PI), David Bennett, Andrew Kusiak, Marian Muste, and Silvia Secchi.

data (CUAHSI 2012a). HIS has pooled datasets from many sources into a coherent and accessible prototype national system for water resource data discovery, delivery, publication, and curation (CUAHSI 2012b). What is missing is the integration of the data into a modeling or forecasting system, which EPA could provide. Problems could be analyzed and solved by using an intelligent digital environmental data system. A human information system is also needed to archive land use, census, voting, planning, and other socioeconomic data relevant to environmental processes and management. The socioeconomic and environmental data would be referenced to common coordinates for use in cross referencing and to enable testing of hypotheses concerning how to solve problems in innovative ways (such as behavioral incentives vs command-and-control).

A central tenet of an intelligent digital environmental data system is that dense, coherent, accessible, multidisciplinary data will serve as an attractor to bring together a broad range of environmental scientists, social scientists, and engineers to pose research questions and devise solutions to environmental problems. It could encourage a social transformation in how interdisciplinary work is accomplished.

Archives and Repositories

It is essential to characterize the environment in diverse ways, although many of the data that result are of limited use without the ability to detect change over time. The implications of exposure to toxic and harmful materials are understood to some extent, but many of the issues being addressed by EPA are in the context of environmental factors whose effects are best characterized in terms of changing exposures or accumulation of materials. Given the great spatial and temporal variability of those same factors, it is often difficult to understand the importance of measurements at a single point in time or space, so measurements of low spatial and temporal scope can easily lead to spurious conclusions. To ensure that EPA and environmental scientists more broadly can effectively understand the relative importance of any single environmental dataset, it is critical to develop and maintain long-term records that are composed of multiple parameters (Lovett et al. 2007). The challenge is to ensure that enough environmental data are collected and preserved to support understanding of long-term trends among the key parameters now identified as important, while providing a high likelihood of providing the data necessary to understand emerging issues. Making data and samples accessible to future researchers are central to ensuring that the understanding of environmental phenomenon continues to grow and evolve with the science. Ensuring that all data collected with federal funds are archived and accessible is critically important, although ideally that would be the norm for all environmental data collected with public or private funds. It is also important to develop sample archives in which materials are appropriately stored for analysis or reanalysis later. New measurement techniques are constantly emerging and providing useful insights. When it is feasible to do the new analyses with stored samples, and thereby create a long-term record of exposure or change, the resulting insights can be invaluable for understanding the implications of new observations (see Rothamsted 2012 for an example of the information that this type of long-term data potentially could provide).

USING NEW SCIENCE TO DRIVE SAFER TECHNOLOGIES AND PRODUCTS

In addition to using new tools and technologies to address the major challenges identified in Chapter 2, it will be important for EPA to continue to look for new ways of preventing environmental problems before they arise. The tools and technologies for measuring and managing scientific data outlined in this chapter have generally been thought of in the context of refined risk-assessment processes. The use of scientific information for the purposes of risk assessment is focused in large part on detailed and nuanced problem identification—that is, a holistic understanding of causes and mechanisms. Such work is important and valuable in understanding how toxicants and other stressors affect environmental health and ecosystems, and at times is required by statute. However, the focus on problem identification often occurs at the expense of efforts to use scientific tools to develop safer technologies and solutions. Consideration of whether functional, cost-effective, and safer alternative manufacturing processes or materials exist that could reduce or eliminate risks while still stimulating innovation is not often part of the risk-assessment processes undertaken by EPA. Given the changing nature of chemical exposures in the United States, from large point sources to disperse, non-point exposures, the traditional tools of exposure assessment and control will likely be insufficient to prevent exposure to chemicals and it may be more effective to place a greater focus on preventing exposure through design changes. NRC (2009) outlined a framework for risk assessment in which the assessment process is tied to evaluating risk-management options rather than the safety of single hazards.

Defining problems without a comparable effort to find solutions greatly diminishes the value of the agency's applied research efforts and may impede its mission to protect human health and the environment. Furthermore, if EPA's actions lead to a change in technology, chemical, or practice, there is a responsibility to understand alternatives and to support a path forward that is environmentally sound, technically feasible, and economically viable (Tickner 2011). Sarewitz et al. (2010) have proposed the Sustainable Solutions Agenda as an alternative approach to think about sustainability problems in the context of complex systems. As noted in other parts of this report, uncertainty is an inevitable part of decision-making processes surrounding complex risks. The Sustainable Solutions Agenda asks a different set of questions about such problems, from asking whether "x causes y" or leads to an "unacceptable" risk to "given current knowledge of the possibility that x causes y, is there a way to move toward more sustainable practice by reducing or replacing x while preserving some or most of its benefits?" Such a focus on solutions through alternatives assessment processes can support the agency's dual science and engineering goals: protection and innovation.

The Pollution Prevention Act of 1990 established the principle that all EPA environmental protection efforts should be based on the prevention or reduction of pollution at the source. Pollution prevention was viewed as such an important program for the agency that its coordination initially occurred through the administrator's office (EPA 2008b). On the basis of the Pollution Prevention Act's mandates, EPA's Office of Research and Development (ORD) and the Office of Pollution Prevention and Toxics (OPPT) embarked on a wide array of initiatives to develop tools, information sources, technologies, approaches, and initiatives to advance pollution prevention. Those have resulted in making EPA a global leader for the application of science and engineering for prevention. Examples of those initiatives are the Cleaner Technologies Substitutes Assessments, the Green Suppliers Network, and a suite of tools to integrate consideration of pollution prevention into chemical design. With more resources and increased coordination at the highest levels of leadership in the agency, EPA has the ability to enhance its support for safer technologies and products. Some mechanisms through which enhanced support can be accomplished include funding research on safer chemistry; building tools so that designers outside the agency can create safer chemicals, products, and processes; providing simple data integration dashboards that will help companies identify and evaluate safer alternatives to chemicals and materials of concern; and setting up consistent guidelines, frameworks, and metrics for evaluating safer chemicals and products.

EPA has taken global leadership in three fields of innovative solutions-oriented science: pollution prevention, Design for the Environment, and green chemistry and engineering. This suite of programs compromises non-regulatory approaches that protect the environment and human health by designing or redesigning processes and products to reduce the use and release of toxic materials. Green chemistry and engineering focuses on molecular design, Design for the Environment focuses on evaluating the safest chemistries and designs for a particular functional use, and pollution prevention focuses on reducing or eliminating emissions and waste in the manufacturing process. The three programs have evolved and changed over time and are overlapping in many ways, but they address different parts of the production process, from chemical design to the use of chemicals in product design to the application in manufacturing. Despite the overlapping connections, the three programs have not been fully integrated in EPA's administrative structures within or between ORD and OPPT, which may ultimately limit the impact and effectiveness of the programs.

Pollution prevention, Design for the Environment, and green chemistry and engineering share a number of common features. First, they have a strong emphasis on education and assistance. To support the change in mindset from "controlling exposure to hazardous materials" to "preventing generation of hazardous materials" or "reducing the hazards of the materials of commerce", each program has developed educational materials. Technical assistance and tools, methods, and expertise are also provided, and research efforts are initiated through ORD and OPPT. Second, they align environmental protection with economic development. A strong incentive for participation in the programs derives from the potential for economic advantages that result from alternative approaches. Using less material or less toxic materials can reduce costs. Innovative solutions driven by environmental concerns can open new markets. Third, they promote strong partnerships between agencies, industry, nongovernment organizations, and academic institutions. The programs recognize participants from outside the agency, including all the stakeholders in the chemical enterprise, as partners that are needed to implement the changes that the programs promote. The partners bring content expertise, research and development resources, and commercialization pipelines that are essential for implementing change and bringing improved products or processes into the marketplace. Fourth, they provide a mechanism for nimbleness. In emphasizing the search for innovative solutions to specific problems, the programs are nimble. In each case, small supporting efforts within the agency support a framework that harnesses and leverages the efforts of innovators in industry, academe, and nongovernment organizations. And fifth, they are a form of voluntary action. Each program promotes participation through incentives as opposed to regulatory approaches. Self-interest of the participants rewards and reinforces participation. These programs are described in greater detail below.

Pollution Prevention

Launched in 1990 through the Pollution Prevention Act (EPA 2011c), EPA's pollution-prevention program was a paradigm shift for the agency in its focus on preventing the generation of waste (source reduction) as opposed to the previous command-and-control, "end-of-pipe" solutions (Browner 1993). The agency recognized that the new approach could be more cost-effective and provide competitive advantages for companies that adopted it. EPA's pollution-prevention efforts have focused on partnerships for prevention in various sectors, such as the automobile, electronics, and health-care sectors; technical support through a network of federal and state technical-assistance providers; technology-development projects; demonstration projects to evaluate technologies; sustainable procurement; and the development of tools for evaluating pollution-prevention options. EPA has established a 2011–2014 strategic plan for pollution prevention that outlines directions for the program (EPA 2010). It has a responsibility to ensure that the market is moving in the right direction, and this can be accomplished through some of the mechanisms described above. Box 3-2 shows an example of how the private sector has influenced the market without the use of regulatory mandates.

> **BOX 3-2** Example of Private Industry's Influence on the Supply Chain without Regulatory Mandates
>
> The increasingly global nature of production, coupled with the expanding number of chemicals used in commerce, presents a daunting challenge for protection of human health and ecosystem quality. It is also challenging for either regulations or the underlying science to keep pace. When the complex interactions between chemicals and their vast production networks are considered, the problems become even more daunting inasmuch as they span organizational and national boundaries.
>
> Partly as a response to those challenges, corporations in several industries have begun to issue supply-chain mandates in which they demand changes in production processes and material inputs from suppliers over which they exert economic influence. Under those mandates, firms in a corporation's supply chain are obliged to meet customer expectations and adopt specific requirements with the promise of future contracts or under the threat of discontinuation of business. The private-sector policies are rapidly emerging as part of a new generation of quasiregulatory policy tools whereby private organizations use economic leverage to effect changes in pursuit of the public good. The efficacy of such mandates is still an open question, but their emergence signals an important evolution and opportunity in the development of strategies aimed at improving consumer protections against exposure to harmful substances. Although such private-sector actions are not a substitute for effective chemical regulations, if developed and executed correctly, they can augment public policies substantially.
>
> An example of the potential influence of managing supply chains is Wal-Mart's commitment to selling high-efficiency light bulbs and the effects that policy had on energy use in the United States (Barbaro 2007). By working with its suppliers on product quality and price, they fulfilled a commitment to quadruple sales of efficient light bulbs. These sales were equal to the maximum sustainable sales—that is, the number of light bulbs per household and the technology available. This effort radically reduced the electricity used by their customers, reducing demand for electricity that was equal to the output of three or four 700 MW power plants.
>
> Another example that demonstrates the power of the marketplace for chemical substitution is bisphenol-A in polycarbonate water bottles (Tickner 2011). As a result of consumer campaigns, emerging science, and state regulations, major retailers of water bottles, such as REI, rapidly switched from polycarbonate to alternative materials. However, in the switch, little research was undertaken on the alternative materials, which has the potential to lead to health and environmental concerns at a later time.

Design for the Environment

Established in 1992, EPA's Design for Environment program is a model of stakeholder-engaged product design to reduce the environmental effects of consumer products. The Design for Environment concept "encourages businesses to incorporate environmental and health considerations in the design and redesign of products and processes" (EPA 2001). It merges several non-regulatory, voluntary initiatives related to the synthesis of chemicals that are safer, an analysis of the risks related to these chemicals, and the development of alternative chemicals and technologies (EPA 2012e). It promotes a collaborative process to improve product design, provides information and tools on design strategies and alternative ingredients, and uses technical assistance, design methods, and a labeling program to create incentives for participation. To achieve its goals, Design for Environment has undertaken cutting-edge research on tools and approaches for advancing safer product design, undertaken a number of supply chain partnerships on more sustainable materials, and engaged in significant outreach with industry and other partners (EPA 1999). Design for the Environment partnerships consider human health and environmental implications, the performance of products, and the economic effects of traditional and alternative chemicals, materials, technologies, and processes (EPA 2006b). In recent years, a primary goal of the Design for the Environment program has been to achieve "informed substitution", that is, moving from a chemical that raises health or environmental concerns to chemicals that are known to be safer or to nonchemical alternatives (EPA 2009). According to EPA (2009), "the goals of informed substitution are to minimize the likelihood of unintended consequences, which can result from a precautionary switch away from a chemical of concern without fully understanding the profile of potential alternatives, and to enable a course of action based on the best information—on the environment and human health—that is available or can be estimated." Design for the Environment achieves its goals through both alternatives-assessment processes and recognition programs. Through its alternatives-assessment processes, Design for the Environment has evaluated alternatives to polybrominated diphenylethers in furniture flame retardants, tetrabromobisphenol-A in printed-circuit boards, and bisphenol-A in thermal cash-register tape, and is examining alternatives to phthalates in wire and cable. The Design for the Environment Safer Product Labeling Program evaluates products and labels the ones that meet the program's safety standards (EPA 2012f). EPA establishes minimum toxicologic criteria for individual cleaning-formulation constituents, and manufacturers submit alternatives that meet the criteria for third-party certification, creating a marketplace for alternative formulations. EPA has developed detailed transparent criteria for evaluation for both programs.

Green Chemistry and Engineering

Green chemistry is another innovative approach to environmental protection that emerged from the Pollution Prevention Act of 1990 (EPA 2011d). Green chemistry and engineering incorporates hazard reduction and waste minimization into design at the molecular level to reduce hazards throughout the life cycle. Green chemistry has been defined as "the utilization of a set of principles that reduces or eliminates the use or generation of hazardous substances in

Emerging Science & Technologies to Address Environmental Challenges   93

the design, manufacture and application of chemical products" and is guided by 12 design principles set forth by Anastas and Warner (2000). The approach goes well beyond reduction of waste and process optimization—it drives the redesign of production and processes at the molecular level. The scope of green chemistry extends beyond alternative-chemicals assessment; when alternatives are not available or suitable, green chemistry principles can be used to design new substances that have environmental performance incorporated at the design stage. The principles address effects throughout the life cycle of a material or product and spur new solutions that represent system-wide reduction in effects as opposed to reduction in effects in only one facet of the life cycle.

Case studies have been used to show how green chemistry and engineering reduce costs and spur innovation by exploring entirely new approaches that are driven by life-cycle thinking and systems thinking. EPA's green chemistry efforts center around research on innovative technologies, provision of tools to evaluate and design green chemistries and processes, and recognition of leadership in green chemistry innovation. For example, the Presidential Green Chemistry Challenge Awards Program (established in 1995) has been used to reward success in green chemistry and to communicate the value of the approach for reducing effects and advancing commercial interests. According to EPA (EPA 2011e),

> During the program's life, EPA has received more than 1,400 nominations and presented awards to 82 winners. Winning technologies alone are responsible for reducing the use or generation of more than 199 million pounds of hazardous chemicals, saving 21 billion gallons of water, and eliminating 57 million pounds of carbon dioxide releases to the air. These benefits are in addition to significant energy and cost savings by the winners and their customers.

Future Implications for Innovation

The three programs described above demonstrate the potential for innovative approaches to advance and use scientific knowledge to protect health and the environment through the redesign of chemicals, materials, and products. They also show the role that EPA can play in driving decisions by providing high-quality scientific information. Since their inception, the three programs have had important data needs and strong links to data generation and tools development. For example, pollution-prevention efforts have been measured using data on chemical release, waste, and use generated by the Toxics Release Inventory and in accordance with several state-level pollution-prevention bills. EPA's Sustainable Futures Initiative has relied on data from EPA's New Chemicals Program and EPA's expertise in chemical assessment and structure–activity relationships to develop tools to assist chemical designers in developing safer products (EPA 2012g). Similarly, efforts are underway to utilize data developed through EPA's ToxCast and ExpoCast programs for the development and evaluation of options for green chemistry and Design for the Environment. A 2011 workshop hosted by the NRC Committee on Emerging Science for Environmental Health Decisions specifically explored applying 21st century toxicology to green chemical and material design (NRC 2011).

The science and engineering tools and technologies for measuring, monitoring, and managing environmental and health data outlined in Chapter 3 can provide essential information to drive sustainable solutions through prevention programs. For example, data on chemicals used in media and sensing can be important in setting priorities among chemicals, processes, and products for prevention actions and for measuring results of such actions; toxicogenomics and exposure data are critical for supporting design and evaluation of new technologies, comparing alternatives throughout their life cycles, and helping to avoid unintended consequences; and crowdsourcing and social-media tools provide a mechanism for sharing information about successful innovations and enhancing existing technical support and demonstration efforts. In addition to traditional environmental sciences, there is a critical need for behavioral and social sciences in advancing the development and adoption of safer chemicals, materials, and products. The data that these scientific disciplines provide are important inputs for characterizing and making the economic case for new technologies, for understanding business and consumer behavior, and for effecting the behavioral changes necessary to ensure such innovations take root in such a way that consumer preferences recognize safer materials.

SUMMARY

In today's information age, explosive amounts of data are generated through all kinds of media, for different purposes, and by commercial or research organizations in both the private and public sectors. It will be a cornerstone of the future of science in general, and EPA's future science in particular, that EPA be able to harvest and synthesize the large amounts of data that transcend geopolitical and scientific disciplinary boundaries. Taking advantage of these data requires a variety of techniques, led by careful problem formulation to ensure that the appropriate data are being collected or analyzed. It also requires state-of-the-art capability in data integration and synthesis, particularly in the areas of data-mining, and in modeling of biologic systems with biostatistics, computer simulation, and other emerging methods. Although the committee notes that it is imperative that EPA be conversant in the latest tools and technologies, a subset of which are discussed in this chapter, it also recognizes that there are substantial constraints on resources. In many cases, building capacity in new and emerging technologies can be achieved through strategic collaborations and should not come at the expense of core disciplines relevant to its mission. These core disciplines include, but are not limited to, statistics, chemistry, economics, environmental engineering, ecology, toxicology, epidemiology, exposure science, and risk assessment. Regardless, leveraging insights from new tools and technologies will be necessary to address some of the emerging problems of the 21st century.

REFERENCES

Aboytes, R., G.D. Di Giovanni, F.A. Abrams, C. Rheinecker, W. Mcelroy, N. Shaw, and M.W. Lechevallier. 2004. Detection of infectious *Cryptosporidium* in filtered drinking water. J. Am. Water Works Assoc. 96(9):88-98. Appel, K.W., S.J. Roselle, R.C. Gilliam, and J.E. Pleim. 2010. Sensitivity of the Community Multiscale Air Quality (CMAQ) model v.4.7 results for the eastern United States to MM5 and WRF meteorological drivers. Geosci. Model Dev. 3(1):169-188.

Ahsan, H., Y. Chen, F. Parvez, M. Argos, A.I. Hussain, H. Momotaj, D. Levy, A. van Geen, G. Howe, and J. Graziano. 2006. Health Effects of Arsenic Longitudinal Study (HEALS): Description of a multidisciplinary epidemiologic investigation. J. Expo. Sci. Environ. Epidemiol. 16:191-205.

Anastas, P.T., and J.C. Warner. 2000. Green Chemistry: Theory and Practice. New York: Oxford University Press.

Appel, K.W., S.J. Roselle, R.C. Gilliam, and J.E. Pleim. 2010. Sensitivity of the Community Multiscale Air Quality (CMAQ) model v4.7 results for the eastern United States to MM5 and WRF meteorological drivers. Geoscientific Model Development. 3(1):169-188.

ASCE (American Society of Civil Engineers). 2004. Interim Voluntary Guidelines for Designing an Online Contaminant Monitoring System. American Society of Civil Engineers, Reston, VA [online]. Available: http://www.michigan.gov/documents/deq/deq-wb-wws-asceocms_265136_7.pdf [accessed Apr. 5, 2012].

Aw, T.G., and J.B. Rose. 2012. Detection of pathogens in water: From phylochips to qPCR to pyrosequencing. Curr. Opin. Biotechnol. 23(3):422-430.

Baker, D., and M.J. Nieuwenhuijsen. 2008. Environmental Epidemiology: Study Methods and Application. Oxford: Oxford University Press.

Barbaro, M. 2007. Wal-Mart puts some muscle behind power-sipping bulbs. January 2, 2007. New York Times [online]. Available: http://www.nytimes.com/2007/01/02/business/02bulb.html?pagewanted=all [accessed Mary 17, 2012].

Bergen, K.M., S.J. Goetz, R.O. Dubayah, G.M. Henebry, C.T. Hunsaker, M.L. Imhoff, R.F. Nelson, G.G. Parker, and V.C. Radeloff. 2009. Remote sensing of vegetation 3-D structure for biodiversity and habitat: Review and implications for LiDAR and radar spaceborne missions. J. Geophys. Res. Biogeosci. 114:G00E06, doi:10.1029/2008JG000883.

Bibby, K., E. Viau, and J. Peccia. 2010. Pyrosequencing of the 16S rRNA gene to reveal bacterial pathogen diversity in biosolids. Water Res. 44(14):4252-4260.

Bibby, K., E. Viau, and J. Peccia. 2011. Viral metagenome analysis to guide human pathogen monitoring in environmental samples. Lett. Appl. Microbiol. 52(4):386-392.

Bierman, P., M. Lewis, B. Ostendorf, and J. Tanner. 2011. A review of methods for analysing spatial and temporal patterns in coastal water quality. Ecol. Indic. 11(1):103-114.

Blaauboer, B.J. 2010. Biokinetic modeling and in vitro-in vivo extrapolations. J. Toxicol. Environ. Health B Crit. Rev. 13(2-4):242-252.

Browner, C.M. 1993. Pollution Prevention Takes Center State. EPA Journal, July/September 1993 [online]. Available: http://www.epa.gov/aboutepa/history/topics/ppa/01.html [accessed Apr. 9, 2012].

Byappanahalli, M.N., R.L. Whitman, D.A. Shively, and M.B. Nevers. 2010. Linking nonculturable (qPCR) and culturable enterococci densities with hydrometeorological conditions. Sci. Total Environ. 408(16):3096-3101.

CAMRA. 2012. Welcome to the CAMRA Microbial Risk Assessment Wiki (CAMRAwiki) [online]. Available: http://wiki.camra.msu.edu/index.php?title=Main_Page [accessed Apr. 10, 2012].

Cangelosi, G.A., K.M. Weigel, C. Lefthand-Begay, and J.S. Meschke. 2010. Molecular detection of viable bacterial pathogens in water by ratiometric pre-rRNA analysis. Appl. Environ. Microbiol. 76(3):960-962.

CDC (Centers for Disease Control and Prevention). 2012. National Report on Human Exposure to Environmental Chemicals. Centers for Disease Control and Prevention [online]. Available: http://www.cdc.gov/exposurereport/ [accessed Mar. 30, 2012].

Christensen, V.G., P.P. Rasmussen, and A.C. Ziegler. 2002. Real-time water quality monitoring and regression analysis to estimate nutrient and bacteria concentrations in Kansas streams. Water Sci. Technol. 45(9):205-211.

Cooper, S., F. Khatib, A. Treuille, J. Barbero, J. Lee, M. Beenen, A. Leaver-Fay, D. Baker, Z. Popović, and Foldit Players. 2010a. Predicting protein structures with a multiplayer online game. Nature 466(7307):756-760.

Cooper, S., A. Treuille, J. Barbero, A. Leaver-Fay, K. Tuite, F. Khatib, A.C. Snyder, M. Beenen, D. Salesin, D. Baker, Z. Popović, and Foldit Players. 2010b. The challenge of designing scientific discovery games. Pp. 40-47 in Proceedings of the Fifth International Conference on the Foundations of Digital Games (FDG 2010), June 19-21, Monterey, CA. New York: ACM [online]. Available: http://www.cs.washington.edu/homes/zoran/foldit-fdg10.pdf [accessed Mar. 30, 2012].

Cooper, S., F. Khatib, I. Makedon, H. Lu, J. Barbero, D. Baker, J. Fogarty, Z. Popović, and Foldit Players. 2011. Analysis of social gameplay macros in the Foldit cookbook. Pp. 9-14 in Proceedings of the Sixth International Conference on the Foundations of Digital Games (FDG 2011), June 28-July 1, 2011, Bordeaux, France. New York: ACM [online]. Available: http://grail.cs.washington.edu/projects/protein-game/foldit-fdg11.pdf [accessed Mar. 30, 2012].

Coppus, R., and A.C. Imeson. 2002. Extreme events controlling erosion and sediment transport in a semi-arid sub-Andean valley. Earth Surf. Proc. Land. 27(13):1365-1375.

Crouse, D.L., P.A. Peters, A. van Donkelaar, M.S. Goldberg, P.J. Villeneuve, O. Bioron, S. Khan, D.O. Atari, J. Jerrett, C.A. Pope, M. Brauer, J.R. Brook, R.V. Martin, D. Stieb, and R.T. Burnett. 2012. Risk of nonaccidental and cardiovascular mortality in relation to long-term exposure to low concentrations of fine particular matter: A Canadian national-level cohort study. Environ. Health Perspect. 120(5):708-714.

CUAHSI (Consortium of Universities for the Advancement of Hydrologic Science, Inc.). 2012a Hydrologic Information System [online]. Available: http://his.cuahsi.org [accessed Apr. 23, 2012].

CUAHSI (Consortium of Universities for the Advancement of Hydrologic Science, Inc.). 2012b. CUAHSI – Data Access, Publication and Analysis [online]. Available: http://www.cuahsi.org/his.html [accessed Apr. 23, 2012].

Diez Roux, A.V. 2011. Complex systems thinking and current impasses in health disparities research. Am. J. Public Health 101(9):1627-1634.

Djikeng, A., R. Kuzmickas, N.G. Anderson, and D.J. Spiro. 2009. Metagenomic analysis of RNA viruses in a fresh water lake. PLoS ONE 4(9):e7264.

Dominici, F., R.D. Peng, C.D. Barr, and M.L. Bell. 2010. Protecting human health from air pollution: Shifting from a single-pollutant to a multipollutant approach. Epidemiology 21(2):187-194.

Eder, B., D. Kang, R. Mathur, J. Pleim, S. Yu, T.A. Otte, and G. Pouliot. 2009. Performance evaluation of the National Air Quality Forecast Capability for the summer of 2007. Atmos. Environ. 43(14): 2312-2320.

Egeghy, P.P., R. Judson, S. Gangwal, S. Mosher, D. Smith, J. Vail, and E.A. Cohen Hubal. 2012. The exposure data landscape for manufactured chemicals. Sci. Total Environ. 414(1):159-166.

Elliott, P., and T.C. Peakman. 2008. The UK Biobank sample handling and storage protocol for the collection, processing and archiving of human blood and urine. Int. J. Epidemiol. 37(2):234-244.

Environmental Information Exchange Network. 2011. Exchange Network Leadership Council [online]. Available: http://www.exchangenetwork.net/about/network-management/exchange-network-leadership-council/ [accessed Apr. 2, 2012].

EPA (US Environmental Protection Agency). 1984. Risk Assessment and Management: Framework for Decision Making. EPA 600/9-85-002. US Environmental Protection Agency, Washington, DC.

EPA (US Environmental Protection Agency). 1986. Ambient Water Quality Criteria for Bacteria-1986: Bacteriological Water Quality Criteria for Marine and Fresh Recreational Waters. EPA-440/5-84-002. Office of Water Regulations and Standards, US Environmental Protection Agency, Washington, DC. January 1986 [online]. Available: http://water.epa.gov/action/advisories/drinking/upload/2009_04_13_beaches_1986crit.pdf [accessed Apr. 3, 2012].

EPA (US Environmental Protection Agency). 1997. Approval of EPA Method 1613 for Analysis of Dioxins and Furans in Wastewater. Fact Sheet, July 1997. Office of Water, US Environmental Protection Agency [online]. Available: http://water.epa.gov/scitech/methods/cwa/organics/dioxins/1613-fs.cfm [accessed Apr. 2, 2012].

EPA (US Environmental Protection Agency). 1999. Design for the Environment: Building Partnerships for Environmental Improvement. EPA744-R-99-003. Office of Pollution Prevention and Toxics, US Environmental Protection Agency, Washington, DC.

EPA (US Environmental Protection Agency). 2000. Risk Characterization Handbook. EPA 100-B-00-002. Office of Science Policy, Office of Research and Development, US Environmental Protection Agency, Washington, DC [online]. Available: http://www.epa.gov/stpc/pdfs/rchandbk.pdf [accessed Aug. 1, 2012].

EPA (US Environmental Protection Agency). 2001. The US Environmental Protection Agency's Design for Environment Program. EPA744-F-00-020. Office of Pollution Prevention and Toxics, US Environmental Protection Agency, Washington, DC. May 2001 [online]. Available: http://www.epa.gov/dfe/pubs/tools/DfEBrochure.pdf [accessed Apr. 11, 2012].

EPA (US Environmental Protection Agency). 2002. Method 1600: Enterococci in Water by Membrane Filtration Using membrane-Enterococcus Indoxyl-β-D-Glucoside Agar (mEI). EPA-821-R-02-022. Office of Water, US Environmental Protection Agency, Washington DC [online]. Available: http://www.caslab.com/EPA-Methods/PDF/EPA-Method-1600.pdf [accessed Apr. 3, 2012].

EPA (US Environmental Protection Agency). 2005. Microbial Source Tracking Guide Document. EPA-600/R-05/064. Office of Research and Development, US Environmental Protection Agency, Washington DC. June 2005 [online]. Available: http://www.ces.purdue.edu/waterquality/resources/MSTGuide.pdf [accessed Apr. 3, 2012].

EPA (US Environmental Protection Agency). 2006a. Implementing the BEACH Act of 2000. Report to Congress. US Environmental Protection Agency [online]. Available: http://water.epa.gov/type/oceb/beaches/report_index.cfm [accessed Aug. 13, 2012].

EPA (US Environmental Protection Agency). 2006b. Design for the Environment Partnership Highlights. US Environmental Protection Agency, March 2006 [online]. Available: http://www.epa.gov/dfe/pubs/about/dfe-highlights06b.pdf [accessed Apr. 11, 2012].

EPA (US Environmental Protection Agency). 2008a. US EPA Office of Research and Development Computational Toxicology Research Program Implementation Plan for Fiscal Years 2009 to 2012: Providing High-Throughput Decision Support Tools for Screening and Assessing Chemical Exposure, Hazard, and Risk. US Environmental Protection Agency [online]. Available: http://epa.gov/ncct/download_files/basic_information/CTRP2_Implementation_Plan_FY09_12.pdf [accessed Apr. 2, 2012].

EPA (US Environmental Protection Agency). 2008b. Evaluation of EPA Efforts to Integrate Pollution Prevention Policy throughout EPA and at Other Federal Agencies. Office of Pollution, Prevention and Toxics, US Environmental Protection Agency. October 2008 [online]. Available: http://www.epa.gov/p2/pubs/docs/p2integration.pdf [accessed Apr. 20, 2012].

EPA (US Environmental Protection Agency). 2009. EPA's DfE Standard for Safer Cleaning Products (SSCP). US Environmental Protection Agency. June 2009 [online]. Available: http://www.epa.gov/opptintr/dfe/pubs/projects/formulat/dfe_criteria_for_cleaning_products_10_09.pdf [accessed Apr. 10, 2012].

EPA (US Environmental Protection Agency). 2010. FY 2011-2015 EPA Strategic Plan: Achieving Our Vision. US Environmental Protection Agency, September 30, 2010 [online]. Available: http://nepis.epa.gov/Adobe/PDF/P1008YOS.PDF [accessed Apr. 2, 2012].

EPA (US Environmental Protection Agency). 2011a. National-Scale Mercury Risk Assessment Supporting the Appropriate and Necessary Finding for Coal- and Oil-Fired Electric Generating Units, Draft. Technical Support Document. EPA-452/D-11-002. Office of Air Quality Planning and Standards, US Environmental Protection Agency, Research Triangle Park, NC. March 2011 [online]. Available: http://yosemite.epa.gov/sab/sabproduct.nsf/02ad90b136fc21ef85256eba00436459/9F048172004D93BB8525783900503486/$File/hg_risk_tsd_3-17-11.pdf [accessed Apr. 4, 2012].

EPA (US Environmental Protection Agency). 2011b. Inventory of US Greenhouse Gas Emissions and Sinks: 1990–2009. EPA 430-R-11-005. US Environmental Protection Agency, Washington, DC. April 15, 2011 [online]. Available: http://www.epa.gov/climatechange/Downloads/ghgemissions/US-GHG-Inventory-2012-Main-Text.pdf [accessed Aug. 2, 2012].

EPA (US Environmental Protection Agency). 2011c. Pollution Prevention (P2): Basic Information. US Environmental Protection Agency [online]. Available: http://www.epa.gov/p2/pubs/basic.htm [accessed Apr. 9, 2012].

EPA (US Environmental Protection Agency). 2011d. Green Chemistry Program at EPA. US Environmental Protection Agency [online]. Available: http://www.epa.gov/greenchemistry/pubs/epa_gc.html [accessed Apr. 9, 2012].

EPA (US Environmental Protection Agency). 2011e. EPA Honors Winners of 2011 Presidential Greem Chemistry Challenge Awards. US Environmental Protection Agency, News: June 20, 2011 [online]. Available: http://yosemite.epa.gov/opa/admpress.nsf/0/93C78AFC58096165852578B5004B1E99 [accessed Apr. 9, 2012].

EPA (US Environmental Protection Agency). 2012a. Computational Toxicology Research, Research Publications. US Environmental Protection Agency [online]. Available: http://www.epa.gov/ncct/publications.html [accessed July 10, 2012].

EPA (US Environmental Protection Agency). 2012b. Quantifying Methane Abatement Efficiency at Three Municipal Solid Waste Landfills. Final Report. EPA/600/R-12/003. Prepared by ARCADIS US, Inc., Durham, NC, for Office of Research and Development, US Environmental Protection Agency, Research Triangle Park, NC. January 2012.

EPA (US Environmental Protection Agency). 2012c. Solid Waste and Emergency Response Discussion Forum. US Environmental Protection Agency [online]. Available: http://blog.epa.gov/oswerforum/ [accessed Apr. 4, 2012].

EPA (US Environmental Protection Agency). 2012d. Watershed Central. Office of Water, US Environmental Protection Agency [online]. Available: http://water.epa.gov/type/watersheds/datait/watershedcentral/index.cfm [accessed Apr. 4, 2012].

EPA (US Environmental Protection Agency). 2012e. Program History. Design for the Environment. US Environmental Protection Agency [online]. Available: http://www.epa.gov/dfe/pubs/about/history.htm [accessed Apr. 9, 2012].

EPA (US Environmental protection Agency). 2012f. Safe Product Labeling. Design for Environment, US Environmental Protection Agency [online]. Available: http://www.epa.gov/opptintr/dfe/pubs/projects/formulat/saferproductlabeling.htm [accessed Apr. 10, 2012].

EPA (US Environmental Protection Agency). 2012g. Sustainable Future. US Environmental Protection Agency [online]. Available: http://www.epa.gov/oppt/sf/ [accessed Apr. 9, 2012].

Ferebee, M., Jr. 2011. New Tools, New Rules, New Year. Inside. Business News, December 2, 2011 [online]. Available: http://insidebiz.com/news/new-tools-new-rules-new-year-whether-its-through-new-technology-processes-behavior-modeling-les [accessed Apr. 5, 2012].

Fishman, J., K.W. Bowman, J.P. Burrows, A. Richter, K.V. Chance, D.P. Edwards, R.V. Martin, G.A. Morris, R. B. Pierce, J.R. Ziemke, J.A. Al-Saadi, J.K. Creilson, T.K. Schaack, and A.M. Thompson. 2008. Remote sensing of tropospheric pollution from space. Bull. Amer. Meteor. Soc. 89(6):805-821.

Foldit. 2012. The Science behind Foldit. Solve the Puzzles for Science [online]. Available: http://fold.it/portal/info/science [accessed Apr. 4, 2012].

Fong, T.T., and E.K. Lipp. 2005. Enteric viruses of humans and animals in aquatic environments: Health risks, detection, and potential water quality assessment tools. Microbiol. Mol. Biol. Rev. 69(2):357-371.

Gilski, M., M. Kazmierczyk, S. Krzywda, H. Zábranská, S. Cooper, Z. Popović, F. Khatib, F. DiMaio, J. Thompson, D. Baker, I. Pichová, and M. Jaskolski. 2011. High-resolution structure of a retroviral protease folded as a monomer. Acta Cryst. D67:907-914.

Goetz, S.J., and R.O. Dubayah. 2011. Advances in remote sensing technology and implications for measuring and monitoring forest carbon stocks and change. Carbon Manag. 2(3):231-244.

Greenbaum, D., and R. Shaikh. 2010. First steps toward multipollutant science for air quality decisions. Epidemiology 21(2):195-197.

Gurney, K.R., R.M. Law, A.S Denning, P.J. Rayner, D. Baker, P. Bousquet, L. Bruhwiler, Y.H. Chen, P. Ciais, S. Fan, I.Y. Fung, M. Gloor, M. Heimann, K. Higuchi, J. John, T. Maki, S. Maksyutov, K. Masarie, P. Peylin, M. Prather, B.C. Pak, J. Randerson, J. Sarmiento, S. Taguchi, T. Takahashi, and C.W. Yuen. 2002. Towards robust regional estimates of CO2 sources and sinks using atmospheric transport models. Nature 415(6872):626-630.

Haas, C.N. 1983. Estimation of risk due to low doses of microorganisms: A comparison of alternative methodologies. Am. J. Epidemiol. 118(4):573-582.

Haas, C.H., J.B. Rose, and C.P. Gerba, eds. 1999. Quantitative Microbial Risk Assessment. New York: John Wiley and Sons.

Hall, F.G., K. Bergen, J.B. Blair, R. Dubayah, R. Houghton, G. Hurtt, J. Kellndorfer, M. Lefsky, J. Ranson, S. Saatchi, H.H. Shugart, and D. Wickland. 2011. Characterizing 3D vegetation structure from space: Mission requirements. Remote Sens. Environ. 115(11):2753-2775.

Hall, J., A.D. Zaffiro, R.B. Marx, , P. C. Kefauver, E.R. Krishnan, R. Haught, and J.G. Herrmann. 2007. On-line water quality parameters as indicators of distribution system contamination. J. Am. Water Works Assoc. 99(1):66-77.

Hayes, D.J., D.P. Turner, G. Stinson, A.D. McGuire, Y. Wei, T.O. West, L.S. Heath, B. de Jong, B.G. McConkey, R.A. Birdsey, W.A. Kurz, A.R. Jacobson, D.N. Huntzinger, Y. Pan, W. Mac Post, and R.B. Cook. 2012. Reconciling estimates of contemporary North American carbon balance among terrestrial biosphere models, atmospheric inversions, and a new approach for estimating net ecosystem exchange from inventory-based data. Global Change Biol. 18(4):1282-1299.

He, S. 2003. Informatics: A brief survey. Electron. Libr. 21(2):117-122.

Heald, C.L., D.J. Jacob, R.J. Park, B. Alexander, T.D. Fairlie, R.M. Yantosca, and D.A. Chu. 2006. Transpacific transport of Asian anthropogenic aerosols and its impact on surface air quality in the United States. J. Geophys. Res. 111:D14310, doi:10.1029/2005JD006847.

Hill, V.R., A.L. Polaczyk, D. Hahn, J. Narayanan, T.L. Cromeans, J.M. Roberts, and J.E. Amburgey. 2005. Development of a rapid method for simultaneous recovery of diverse microbes in drinking water by ultrafiltration with sodium polyphosphate and surfactants. Appl. Environ. Microbiol. 71(11):6878-6884.

Hystad, P., E. Setton, A. Cervantes, K. Poplawski, S. Deschenes, M. Brauer, A. van Donkelaar, L. Lamsal, R. Martin, M. Jerrett, and P. Demers. 2011. Creating national air pollution models for population exposure assessment in Canada. Environ. Health Perspect. 119(8):1123-1129.

Jean, J., Y. Perrodin, C. Pivot, D. Trep, M. Perraud, J. Droguet, F. Tissot-Guerraz, F. Locher. 2012. Identification and prioritization of bioaccumulable pharmaceutical substances discharged in hospital effluents. J. Environ. Manage. 103:113-121.

Jeong, Y., B.F. Sanders, and S.B. Grant. 2006. The information content of high-frequency environmental monitoring data signals pollution events in the coastal ocean. Environ. Sci. Technol. 40(20):6215-6220.

Johns, D.O., L.W. Stanek, K. Walker, S. Benromdhane, B. Hubbell, M. Ross, R.B. Devlin, D.L. Costa, and D. S. Greenbaum. 2012. Practical advancement of multipollutant scientific and risk assessment approaches for ambient air pollution. Environ. Health Perspect. 102(9):1238-1242.

Jones, L., J.D. Parker, and P. Mendola. 2010. Blood Lead and Mercury Levels in Pregnant Women in the United States, 2003-2008. NCHS Data Brief No. 52. US Department of Health & Human Services, Centers for Disease Control and Prevention, National

Center for Health Statistics, Hyattsville, MD. December 2010 [online]. Available: http://www.cdc.gov/nchs/data/databriefs/db52.pdf [accessed Apr. 2, 2012].

Judson, R.S., K.A. Houck, R.J. Kavlock, T.B. Knudsen, M.T. Martin, H.M. Mortensen, D.M. Reif, D.M. Rotroff, I. Shah, A.M. Richard, and D.J. Dix. 2010. In vitro screening of environmental chemicals for targeted testing prioritization: The ToxCast project. Environ Health Perspect. 118(4):485-492.

Judson, R.S., R.J. Kavlock, R.W. Setzer, E.A. Cohen-Hubal, M.T. Martin, T.B. Knudsen, K.A. Houck, R.S. Thomas, B.A. Wetmore, and D.J. Dix. 2011. Estimating toxicity-related biological pathway altering doses for high-throughput chemical risk assessment. Chem. Res. Toxicol. 24(4):451-462.

Kaiser, J. 2012. Overhaul of the US Child Health Study Concerns Investigators. Science. 335:1032.

Khatib, F., S. Cooper, M.D. Tyka, K. Xu, I. Makedon, Z. Popović, D. Baker, and Foldit Players. 2011a. Algorithm discovery by protein folding game players. Proc. Natl. Acad. Sci. 108(47):18949-18953.

Khatib, F., F. DiMaio, Foldit Contenders Group, Foldit Void Crushers Group, S. Cooper, M. Kazmierczyk, M. Gilski, S. Krzywda, H. Zábranská, I. Pichová, J. Thompson, Z. Popović, M. Jaskolski, and D. Baker. 2011b. Crystal structure of a monomeric retroviral protease solved by protein folding game players. Nat. Struct. Mol. Biol. 18(10):1175-1177.

Koetz, B., G.Q. Sun, F. Morsdorf, K.J. Ranson, M. Kneubuhler, K. Itten, and B. Allgower. 2007. Fusion of imaging spectrometer and LiDAR data over combined radiative transfer models for forest canopy characterization. Remote Sens. Environ. 106(4):449-459.

Kristiansson, E., J. Fick, A. Janzon, R. Grabic, C. Rutgersson, B. Weijdegård, H. Söderström, and D.G.J. Larsson. 2011. Pyrosequencing of antibiotic-contaminated river sediments reveals high levels of resistance and gene transfer elements. PLoS ONE 6(2):e17038.

Lioy, P.J., and S.M. Rappaport. 2011. Exposure science and the exposome: An opportunity for coherence in the environmental health sciences. Environ. Health Perspect. 119(11):A466-A467.

Lioy, P.J., S.S. Isukapalli, L. Trasande, L. Thorpe, M. Dellarco, C. Weisel, P.G. Georgopoulos, C. Yung, S. Alimokhtari, M. Brown, and P.J. Landrigan. 2009. Using national and local extant data to characterize environmental exposures in the national children's study: Queens County, New York. Environ. Health Perspect. 117(10):1494-1504.

Long, S.C., and J.D. Plummer. 2004. Assessing land use impacts on water quality using microbial source tracking. J. Am. Water Resour. Assoc. 40(6):1433-1448.

Longley, D., and M. Shain. 1985. Dictionary of Information Technology, 2 Ed. London: Macmillan Press.

Loperfido, J.V., C.L. Just, and J.L. Schnoor. 2009. High-frequency diel dissolved oxygen stream data modeled for variable temperature and scale. J. Environ. Eng.-ASCE 135(12):1250-1256.

Loperfido, J.V., P. Beyer, C.L. Just, and J.L. Schnoor. 2010a. Uses and biases of volunteer water quality data. Environ. Sci. Technol. 44(19):7193-7199.

Loperfido, J.V., C.L. Just, A.N. Papanicolaou, and J.J. Schnoor. 2010b. In situ sensing to understand diel turbidity cycles, suspended solids, and nutrient transport in Clear Creek, Iowa. Water Resour. Res. 46:W06525, doi: 10.1029/2009WR008293.

Lovett, G.M., D.A. Burns, C.T. Driscoll, J.C. Jenkins, M.J. Mitchell, L. Rustad, J.B. Shanley, G.E. Likens, and R. Haeuber. 2007. Who needs environmental monitoring? Front. Ecol. Environ. 5(5):253-260.

Martin, R.V. 2008. Satellite remote sensing of surface air quality. Atmos. Environ. 42(34):7823-7843.

Matthews, M.W. 2011. A current review of empirical procedures of remote sensing in inland and near-coastal transitional waters. Int. J. Remote Sens. 32(21):6855-6899.

McHale, C.M., L. Zhang, A.E. Hubbard, and M.T. Smith. 2010. Toxicogenomic profiling of chemically exposed humans in risk assessment. Mutat. Res. 705(3):172-183.

Medema, G.J., W. Hoogenboezem, A.J. van der Veer, H.A. Ketelaars, W.A. Hijnen, and P.J. Nobel. 2003. Quantitative risk assessment of Cryptosporidium in surface water treatment. Water Sci. Technol. 47(3):241-247.

Mertes, L.A.K. 2002. Remote sensing of riverine landscapes. Freshwater Biol. 47(4):799-816.

Messer, J.W., and A.P. Dufour. 1998. A rapid, specific membrane filtration procedure for enumeration of enterococci in recreational water. Appl. Environ. Microbiol. 64(2):678-680.

Messner, M., S. Shaw, S. Regli, K. Rotert, V. Blank, and J. Soller. 2006. An approach for developing a national estimate of waterborne disease due to drinking water and a national estimate model application. J. Wat. Health 4(suppl. 2):201-240.

Morris, G.A., S. Hersey, A.M. Thompson, S. Pawson, J.E. Nielsen, P.R. Colarco, W.W. McMillan, A. Stohl, S. Turquety, J. Warner, B.J. Johnson, T.L. Kucsera, D.E. Larko, J. Oltmans, and J.C. Witte. 2006. Alaskan and Canadian forest fires exacerbate ozone pollution over Houston, Texas, on 19 and 20 July 2004. J. Geophys. Res. 111:D24S03, doi:10.1029/2006JD007090.

Napelenok, S.L., R.W. Pinder, A.B. Gilliland, and R.V. Martin. 2008. A method for evaluating spatially-resolved NO_x emissions using Kalman filter inversion, direct sensivitities, and space-based NO_2 observations. Atmos. Chem. Phys. 8(18):5603-5614.

Neng, N.R., and J.M. Nogueira. 2012. Development of a bar adsorptive micro-extraction-large-volume injection-gas chromatography-mass spectrometric method for pharmaceuticals and personal care products in environmental water matrices. Anal. Bioanal. Chem. 402(3):1355-1364.

NHLBI (National Heart, Lung and Blood Institute). 2011. NHLBI Population Studies Database [online]. Available: http://apps.nhlbi.nih.gov/popstudies/ [accessed Oct. 11, 2011].

NIH (National Institutes of Health). 2012. Agricultural Health Study [online]. Available: http://aghealth.nci.nih.gov/ [accessed May 3, 2012].

Noble, R.T., A.D. Blackwood, J.F. Griffith, C.D. McGee, and S.B. Weisberg. 2010. Comparison of rapid quantitative PCR-based and conventional culture-based methods for enumeration of Enterococcus spp. and *Escherichia coli* in recreational waters. Appl. Environ. Microbiol. 76(22):7437-7443.

Nowak, P., S. Bowen, and P. Cabot. 2006. Disproportionality as a framework for linking social and biophysical systems. Soc. Natur. Resour. 19(2):153-173.

NRC (National Research Council). 1983. Risk Assessment in Federal Government: Managing the Process. Washington, DC: National Academy Science.

NRC (National Research Council). 2007a. Toxicity Testing in the 21st Century: A Vision and a Strategy. Washington, DC: National Academies Press.

NRC (National Research Council). 2007b. Models in Environmental Regulatory Decision Making. Washington, DC: National Academies Press.

NRC (National Research Council). 2009. Science and Decisions: Advancing Risk Assessment. Washington, DC: National Academies Press.

NRC (National Research Council). 2011. Emerging Science for Environmental Health, Decisions: Workshop: Applying 21st Century Toxicology to Green Chemical and Material Design [online]. Available: http://nas-sites.org/emergingscience/workshops/green-chemistry/workshop-presentations-green-chemistry/ [accessed Apr. 11, 2012].

NRC (National Research Council). 2012. A Research Strategy for Environmental, Health and Safety Aspects of Engineered Nanomaterials. Washington, DC: National Academy Press.

NRC/IOM (National Research Council and Institute of Medicine). 2008. The National Children's Study Research Plan: A Review. Washington, DC: The National Academies Press.

NRDC (National Resource Defense Council). 2011. Testing the Waters: A Guide to Water Quality at Vacation Beaches, 21st Annual Report. National Resource Defense Council [online]. Available: http://www.nrdc.org/water/oceans/ttw/titinx.asp [accessed Apr. 6, 2012].

Ostby, F.P. 1999. Improved accuracy in severe storm forecasting by the Severe Local Storms Unit during the last 25 years: Then versus now. Weather Forecast. 14(4): 526-543.

Paules, R. 2003. Phenotypic anchoring: Linking cause and effect. Environ. Health Perspect. 111(6): A338-A339.

Pepper, I.L., J.P. Brooks, R.G. Sinclair, P.L. Gurian, and C.P. Gerba. 2010. Pathogens and indicators in United States Class B Biosolids: National and historic distributions. J. Environ. Qual. 39(6): 2185-2190.

Pillmann, W., W. Geiger, and K. Voigt. 2006. Survey of environmental informatics in Europe. Environ. Modell. Softw. 21(11):1519-1527.

Preuss, P. 2011. ORD Innovation. Presentation to the Science Advisory Board and Board of Scientific Counselors. June 29-30, 2011 [online]. Available: http://yosemite.epa.gov/sab/sabproduct.nsf/962C02E5D5A1587B852578BC005BD263/$File/CORRECTED+PWP+PPT+SAB-BOSC+for+posting.pdf [accessed April 10, 2012].

PSP (Puget Sound Partnership). 2011. The Action Agenda in South Central Puget Sound, Appendix C. Puget Sound National Estuary Program Management Conference Overview. Draft, December 9, 2011 [online]. Available: http://www.psp.wa.gov/downloads/AA2011/120911/AA-draft-120911-appendixC.pdf [accessed Apr. 3, 2011].

Rappaport, S.M., and M.T. Smith. 2010. Environment and disease risks. Science 330 (6003):460-461.

Riboli, E., K.J. Hunt, N. Slimani, N.T. Ferrari, T. Norat, M. Fahey, U.R. Charrondière, B. Hémon, C. Casagrande, J. Vignat, K. Overvad, A. Tjønneland, F. Clavel-Chapelon, A. Thiébaut, J. Wahrendorf, H. Boeing, D. Trichopoulos, A. Trichopoulou, P. Vineis, D. Palli, H.B. Bueno-De-Mesquita, P.H. Peeters, E. Lund, D. Engeset, C.A. González, A. Barricarte, G. Berglund, G. Hallmans, N.E. Day, T.J. Key, R. Kaaks, and R. Saracci. 2002. European Prospective Investigation into Cancer and Nutrition (EPIC): Study populations and data collection. Public Health Nutr. 5(6B):1113-1124.

Ridolfi, L., P. D'Odorico, A. Porporato, and I. Rodriguez-Iturbe. 2003. Stochastic soil moisture dynamics along a hillslope. J. Hydrol. 272(1-4):264-275.

Rosario, K., C. Nilsson, Y.W. Lim, Y.J. Ruan, and M. Breitbart. 2009. Metagenomic analysis of viruses in reclaimed water. Environ. Microbiol. 11(11):2806-2820.

Rose, J.B. 1988. Occurrence and significance of *Cryptosporidium* in water. J. Am. Water Works Assoc. 80(2):53-58.
Rothamsted. 2012. The Rothamsted Archive. Rothamsted Research, Harpenden, U.K. [online]. Available: http://www.rothamsted.ac.uk/Content.php?Section=Resources&Page=RothamstedArchive [accessed Apr. 3, 2012].
Rotroff, D.M., B.A. Wetmore, D.J. Dix, S.S. Ferguson, H.J. Clewell, K.A. Houck, E.L. Lecluyse, M.E. Andersen, R.S. Judson, C.M. Smith, M.A. Sochaski, R.J. Kavlock, F. Boellmann, M.T. Martin, D.M. Reif, J.F. Wambaugh, and R.S. Thomas. 2010. Incorporating human dosimetry and exposure into high-throughput in vitro toxicity screening. Toxicol Sci. 117(2):348-358.
Sarewitz, D., D. Kriebel, R. Clapp, C. Crumbley, P. Hoppin, M. Jacobs, and J. Tickner. 2010. The Sustainable Solutions Agenda. The Consortium for Science, Policy and Outcomes, Arizona State University, Lowell Center for Sustainable Production, University of Massachusetts Lowell [online]. Available: http://www.sustainableproduction.org/downlods/SSABooklet.pdf [accessed Apr. 10, 2012].
Sauch, J.F. 1985. Use of immunofluorescence and phase-contract microscopy for detection and identification of Gardia cysts in water samples. Appl. Environ. Microbiol. 50(6):1434-1438.
Scavia, D., N.N. Rabalais, R.E. Turner, D. Justic, and W.J. Wiseman. 2003. Predicting the response of Gulf of Mexico hypoxia to variations in Mississippi River nitrogen load. Limnol. Oceanogr. 48(3):951-956.
Schaepman, M.E., S.L. Ustin, A.J. Plaza, T.H. Painter, J. Verrelst, and S.L. Liang. 2009. Earth system science related imaging spectroscopy: An assessment. Remote Sens. Environ. 113(1):S123-S137.
Schober, S.E.,T.H. Sinks, R.L. Jones, P.M. Bolger, M. McDowell, J. Osterloh, E.S. Garrett, R.A. Canady, C.F. Dillon, Y. Sun, C.B. Joseph, and K.R. Mahaffey. 2003. Blood mercury levels in US children and women of childbearing age, 1999-2000. JAMA 289(13):1667-1674.
Schwalm, C.R., C.A. Williams, K. Schaefer, R. Anderson, M.A. Arain, I. Baker, A. Barr, T.A. Black, G. Chen, J.M. Chen, P. Ciais, K.J. Davis, A. Desai, M. Dietze, D. Dragoni, M.L. Fischer, L.B. Flanagan, R. Grant, L. Gu, D. Hollinger, R.C. Izaurralde, C. Kucharik, P. Lafleur, B.E. Law, L. Li, Z. Li, S. Liu, E. Lokupitiya, Y. Luo, S. Ma, H. Margolis, R. Matamala, H. McCoughey, R.K. Monson, W.C. Oechel, C. Peng, B. Poulter, D.T. Price, D.M. Riciutto, W. Riley, A.K. Sahoo, M. Sprintsin, J. Sun, H. Tian, C. Tonitto, H. Verbeeck, and S.B. Verma. 2010. A model-data intercomparison of CO2 exchange across North America: Results from the North American carbon program site synthesis. J. Geophys. Res. 115:G00H05, doi:10.1029/2009JG001229.
Seminara, D., M.J. Khoury, T.R. O'Brien, T. Manolio, M.L. Gwinn, J. Little, J.P. Higgins, J.L. Bernstein, P. Boffetta, M. Bondy, M.S. Bray, P.E. Brenchley, P.A. Buffler, J.P. Casas, A.P. Chokkalingam, J. Danesh, G. Davey Smith, S. Dolan, R. Duncan, N.A. Gruis, M. Hashibe, D. Hunter, M.R. Jarvelin, B. Malmer, D.M. Maraganore, J.A. Newton-Bishop, E. Riboli, G. Salanti, E. Taioli, N. Timpson, A.G. Uitterlinden, P. Vineis, N. Wareham, D.M. Winn, R. Zimmern, and J.P. Ioannidis. 2007. The emergence of networks in human genome epidemiology: Challenges and opportunities. Epidemiology 18(1):1-8.
Shanks, O. 2011. Global Inter-lab Fecal Source Tracking Methods Comparison Study. Presentation at 2011 National Beach Conference, March 16, 2011, Miami, FL.
Sheldon, L.S., and E.A. Cohen Hubal. 2009. Exposure as part of a systems approach for assessing risk. Environ. Health Perspect. 117(8):1181-1184.

Shukla, S., C.Y. Yu, J.D. Hardin, and F.H. Jaber. 2006. Wireless data acquisition and control systems for agricultural water management projects. HortTechnology 16(4):595-604.

Sivapalan, M., K. Takeuchi, S.W. Franks, V.K. Gupta, H. Karambiri, V. Lakshmi, X. Liang, J.J. McDonnell, E.M. Mendiondo, P.E. O'Connell, T. Oki, J.W. Pomeroy, D. Schertzer, S. Uhlenbrook, and E. Zehe. 2003. IAHS decade on Predictions in Ungauged Basins (PUB), 2003–2012: Shaping an exciting future for the hydrological sciences. Hydrol. Sci. 48(6):857-880.

Slifko, T.R., D. Friedman, J.B. Rose, and W. Jakubowski. 1997. An in vitro method for detecting infectious *Cryptosporidium* oocysts with cell culture. Appl. Environ. Microbiol. 63(9):3669-3675.

Slifko, T.R., D.E. Huffman, and J.B. Rose. 1999. A most probable assay for enumeration of infectious *Cryptosporidium parvum* oocysts. Appl. Environ. Microbiol. 65(9): 3936-3941.

Srinivasan, S., A. Aslan, I. Xagoraraki, E. Alocilja, and J.B. Rose. 2011. Escherichia coli, enterococci, and Bacteroides thetaiotaomicron qPCR signals through wastewater and septage treatment. Water. Res. 45(8):2561-2572.

Tapscot, D., and A.D. Willams. 2010. MacroWikinomics - Rebooting Business and the World. New York: Penguin.

Tian, D., D.S. Cohan, S. Napelenok, M. Bergin, Y. Hu, M. Chang, and A.G. Russell. 2010. Uncertainty analysis of ozone formation and response to emission controls using higher-order sensitivities. J. Air Waste Manage. Assoc. 60(7):797-804.

Tickner, J.A. 2011. Science of problems, science of solutions or both? A case example of bisphenol. Am. J. Epidemiol. Community Health 65(8):649-650.

Tornero-Velez, R., P.P. Egeghy, and E.A. Cohen Hubal. 2012. Biogeographical analysis of chemical co-occurrence data to identify priorities for mixtures research. Risk Anal. 32(2):224-236.

Tsow, F., E. Forzani, A. Rai, R. Wang, R. Tsui, S. Mastrioianni, C. Knobbe, A.J. Gandolfi, and N.J. Tao. 2009. A wearable and wireless sensor system for real-time monitoring of toxic environmental volatile organic compounds. 9(12):1734-1740.

University of Washington. 2011. MESA Air Pollution [online]. Available: http://depts.washington.edu/mesaair/ [accessed Apr. 3, 2012].

van Donkelaar, A., R.V. Martin, M. Brauer, R. Kahn, R. Levy, C. Verduzco, and P.J. Villeneuve. 2010. Global estimates of ambient fine particulate matter concentrations from satellite-based aerosol optical depth: Development and application. Environ. Health Perspect. 118(6):847-855.

Vandenberghe, V., P.L. Goethals, A. Van Griensven, J. Meirlaen, N. De Pauw, P. Vanrolleghem, and W. Bauwens. 2005. Application of automated measurement stations for continuous water quality monitoring of the Dender river in Flanders, Belgium. Environ. Monit. Assess. 108(1-3):85-98.

Varma, M., R. Field, M. Stinson, B. Rukovets, L. Wymer, and R. Hauglanda. 2009. Quantitative real-time PCR analysis of total and propidium monoazide-resistant fecal indicator bacteria in wastewater. Water Res. 43(19):4790-4801.

Wade, T.J., R.L. Calderon, E. Sams, M. Beach, K.P. Brenner, A.H. Williams, and A.P. Dufour. 2006. Rapidly measured indicators of recreational water quality are predictive of swimming-associated gastrointestional illness. Environ. Health Perspect. 114(1):24-28.

Wang, K., S.E. Franklin, X. Guo, and M. Cattet. 2010. Remote sensing of ecology, biodiversity and conservation: A review from the perspective of remote sensing specialists. Sensors 10(11):9647-9667.

WATERS Network. 2009. Living in the Water Environment: The WATERS Network Science Plan, May 15, 2009 [online]. Available: http://www.watersnet.org/docs/WATERS_Network_SciencePlan_2009May15.pdf [accessed Aug. 8, 2012].

Weis, B.K., D. Balshaw, J.R. Barr, D. Brown, M. Ellisman, P. Lioy, G. Omenn, J.D. Potter, M.T. Smith, L. Sohn, W.A. Suk, S. Sumner, J. Swenberg, D.R. Walt, S. Watkins, C. Thompson, and S.H. Wilson. 2005. Personalized exposure assessment: Promising approaches for human environmental health research. Environ Health Perspect 113(7):840-848.

Wetmore, B.A., J.F. Wambaugh, S.S. Ferguson, M.A. Sochaski, D.M. Rotroff, K. Freeman, H.J. Clewell, III, D.J. Dix, M.E. Andersen, K.A. Houck, B. Allen, R.S. Judson, R. Singh, R.J. Kavlock, A.M. Richard, and R.S. Thomas. 2012. Integration of dosimetry, exposure, and high-throughput screening data in chemical toxicity assessment. Toxicol. Sci. 125(1):157-174.

WHO (World Health Organization). 2004. Guidelines for Drinking-Water Quality: Volume 1 Recommendations, 3rd Ed. Geneva: World Health Organization [online]. Available: http://www.who.int/water_sanitation_health/dwq/GDWQ2004web.pdf [accessed Aug. 1, 2012].

Wild, C.P. 2005. Complementing the genome with an "exposome": The outstanding challenge of environmental exposure measurement in molecular epidemiology. Cancer Epidemiol. Biomarkers Prev. 14(8):1847-1850

Willett, W.C., W.J. Blot, G.A. Colditz, A.R. Folsom, B.E. Henderson, and M.J. Stampfer. 2007. Merging and emerging cohorts: Not worth the wait. Nature 445(7125):257-258.

Yates, M.V., J. Malley, P. Rochelle, and R. Hoffman. 2006. Effect of adenovirus resistance on UV disinfection requirements: A report on the state of adenovirus science. J. Am. Water Works Assoc. 98(6):93-106.

Ye, L., and T. Zhang. 2011. Pathogenic bacteria in sewage treatment plants as revealed by 454 pyrosequencing. Environ. Sci. Technol. 45(17):7173-7179.

Zegura, B., E. Heath, A. Cernosa, and M. Filipic. 2009. Combination of in vitro bioassays for the determination of cytotoxic and genotoxic potential of wastewater, surface water and drinking water samples. Chemosphere. 75(11):1453-1460.

4

Building Science for Environmental Protection in the 21st Century

Since its formation in 1970, the US Environmental Protection Agency (EPA) has had a leadership role in developing the many fields of environmental science and engineering. From ecology to health sciences, environmental engineering to analytic chemistry, EPA has stimulated and supported academic research, developed environmental education programs, supported regional science initiatives, supported and promoted the development of safer and more cost-effective technologies, and provided a firm scientific basis of regulatory decisions and prepared the agency to address emerging environmental problems. The broad reach of EPA science has also influenced international policies and guided state and local actions. The nation has made great progress in addressing environmental challenges and improving environmental quality in the 40 years since the first Earth Day.

As a regulatory agency, EPA applies many of its resources to implementing complex regulatory statutes, including substantial commitments of scientific and technical resources to environmental monitoring, applied health and environmental science and engineering, risk assessment, benefit–cost analysis, and other activities that form the foundation of regulatory actions. The primary focus on its regulatory mission can engender controversy and place strains on the conduct of EPA's scientific work in ways that do not affect most other government science agencies (such as the National Institute of Environmental Health Sciences and the National Science Foundation). Amid this inherent tension, research in EPA generally, and in the Office of Research and Development (ORD) in particular, strives to meet the following objectives:

- Support the needs of the agency's present regulatory mandates and timetables.
- Identify and lay the intellectual foundations that will allow the agency to meet environmental challenges that it faces and will face over the course of the next several decades.

- Determine the main environmental research problems on the US environmental-research landscape.
- Sustain and continually rejuvenate a diverse inhouse scientific research staff—with the necessary laboratories and field capabilities—that can support the agency in its present and future missions and in its active collaboration with other agencies.
- Strike a balance between inhouse and extramural research investment. The latter can often bring new ideas and methods to the agency, stimulate a flow of new people into it, and support the continued health of environmental research in the nation.

Those multiple objectives can lead to conflict. For example, ORD resources that are applied to expanding staff and expediting science reviews and risk assessment in the National Center for Environmental Assessment may divert resources from longer-term program development and research. However, the agency has shown itself capable of maintaining a longer-term perspective in several instances, such as the establishment and maintenance of the Science to Achieve Results (STAR) grant program for extramural research, anticipatory moves to develop capability in computational toxicology, and the development and sustained implementation of multiyear research plans, for example, for research on airborne particulate matter (now the Air Quality, Climate, and Energy multiyear plan). In each of those cases, EPA identified ways both to give longer-term goals higher priority and to identify and commit resources to them. However, the tension between the near-term and longer-term science goals for the agency is likely to increase as more and more contentious rules are brought forward and as continuing budget pressures constrain and reduce science resources overall.

In light of the inherent tension, the emerging environmental issues and challenges identified in Chapter 2, and the emerging science and technologies described in Chapter 3, this chapter attempts to identify key strategies for building science for environmental protection in the 21st century in EPA and beyond. Specifically, the chapter lays out a path for EPA to retain and expand its leadership in science and engineering by establishing a 21st century framework that embraces systems thinking to produce science to inform decisions. That path includes staying at the leading edge by engaging in science that anticipates, innovates, is long term, and is collaborative; using enhanced systems-analysis tools and expertise; and using synthesis research to support decisions. In supporting environmental science and engineering for the 21st century, EPA will need to continue to evolve from an agency that focuses on using science to characterize risks so that it can respond to problems to an agency that applies science to anticipate and characterize both problems and solutions at the earliest point possible. Anticipating and characterizing problems and solutions should optimize social, economic, and environmental factors.

EMBRACING SYSTEMS THINKING FOR PRODUCING AND APPLYING SCIENCE FOR DECISIONS: A 21ST CENTURY FRAMEWORK FOR SCIENCE TO INFORM DECISIONS

The continued emergence of major new and complex challenges described in Chapter 2—and the need to deal with the inevitable uncertainty that accompanies major environmental, technologic, and health issues—will necessitate a new way to make decisions. As described in Chapter 3, *systems thinking* has begun to take root in biology and other fields as a means of considering the whole rather than the sum of its parts; this will be essential as increasingly complex problems and the challenges described in Chapter 2 present themselves. The emergence of "wicked problems", the increasing need to address exposures of humans and the ecosystem to multiple pollutants through multiple pathways (some of which are global), and the continuing challenges for the analysis and characterization of uncertainty throughout science and decision-making combine to make the adoption of systems thinking critical.

The systems-thinking perspective is useful not only for characterizing complex effects but for designing sustainable solutions, whether they are innovative technologies or behavioral changes. Understanding systems is also important for determining where leverage points exist for the prevention of health and environmental effects (Meadows 1999). To successfully inform future environmental protection decisions in an increasingly complex world, systems thinking must, at a minimum, include consideration of cumulative effects of multiple stressors, evaluation of a wide range of alternatives to the activity of concern, analysis of the upstream and downstream life-cycle implications of current and alternative activities, involvement of a broad range of stakeholders in decisions (particularly where uncertainty is significant), and use of interdisciplinary scientific approaches that characterize and communicate uncertainties as clearly as possible. As part of a systems perspective, it will be important for the agency to engage in "systems mapping" to comprehensively understand the way in which interacting stressors (such as environmental, human, technologic, socioeconomic, and political stressors) map to health and environmental impacts and to identify where intervention points can result in primary prevention solutions.

Although EPA has made efforts over the years to attempt to bring systems concepts into its work, most recently in its efforts to reorganize its activities under a sustainability framework (Anastas 2012), these efforts have rarely been integrated throughout the agency, nor sustained from one set of leaders to another. To begin to address the lack of a sustained systems perspective, the committee has developed a 21st century framework for decisions (Figure 4-1) and recommends a set of organizational changes to implement that framework (see Chapter 5). The framework features four elements that will be critical for informing the complex decisions that EPA faces:

- To stay at the leading edge, EPA science will need to
    - Anticipate.

o Innovate.
  o Take the long view.
  o Be collaborative.

- EPA will need to continue to evaluate and apply the new tools for data acquisition, modeling, and knowledge development described in Chapter 3.
- EPA will need to continue to develop and apply new systems-level tools and expertise for systematic analysis of the health, environmental, social, and economic implications of individual decisions.
- EPA will need to continue to develop tools and methods for synthesizing science and characterizing uncertainties, and will need to integrate methods for tracking and assessing the outcomes of actions (that is, for being accountable) into its decision process from the outset.

STAYING AT THE LEADING EDGE OF SCIENCE

EPA can maintain its global position in environmental protection by staying at the leading edge of science and engineering research. Staying at the edge of science knowledge requires staying at the edge of science practice. In addition to understanding the latest advances in the science and practice of environmental protection, EPA will need to continue to engage actively in the identification of emerging scientific and technologic developments, respond to advances in science and technology, and use its knowledge, capacity, and experience to direct those advances. That is consistent with the two principal goals for science in the agency: to safeguard human health and the environment and to foster the development and use of innovative technologies (EPA 2012).

For EPA to stay at the leading edge, the committee presents a set of overarching principles for research and policy that begins to address the challenges of wicked problems. To be able to predict and adequately address existing challenges and prevent on-the-horizon challenges, EPA's science will need to

- *Anticipate.* Be deliberate and systematic in anticipating scientific, technology, and regulatory challenges.
- *Innovate.* Support innovation in scientific approaches to characterize and prevent problems and to support solutions through more sustainable technologies and practices.
- *Take the long view.* Track progress in ecosystem quality and human health over the medium term and the long term and identify needs for midcourse corrections.
- *Be collaborative.* Support interdisciplinary collaboration in and outside the agency, across the United States, and globally.

Those four principles support the flow of science information (from data to knowledge) in EPA to inform environmental decision-making and strategies for inducing desirable environmental behaviors.

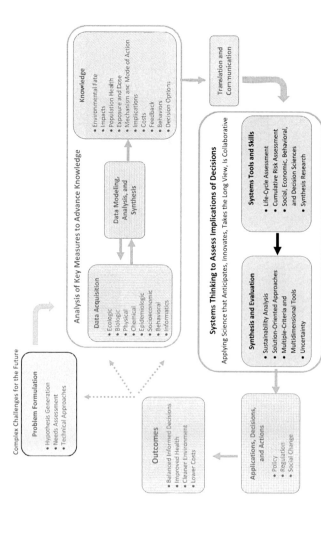

FIGURE 4-1 The iterative process of science-informed environmental decision-making and policy. Leading-edge science will produce large amounts of new information about the state of human health and ecologic systems and the likely effects of introducing a variety of pollutants or other perturbations into the systems. In particular, many multifactorial problems require systems thinking that can be readily integrated into other analytic approaches. This framework relies on science that anticipates, innovates, takes the long view, and is collaborative to solve environmental and human health problems. It also supports decision-making and ensures that leading-edge science is developed and applied to inform assessments of the system-wide implications of alternatives for key policy decisions.

Science That Anticipates

Continually striving to more effectively anticipate challenges and emerging environmental issues will help EPA to stay at the leading edge of science. That involves two main sets of activities: anticipating concerns and developing guidance to avoid problems with new or emerging technologies, and establishing key indicators and tracking trends in human health and ecosystem quality to identify and dedicate resources to emerging environmental problems. Furthermore, continuing to anticipate (and direct resources to) targeted science and technology developments will allow EPA to enhance its ability to identify early warnings and prevent effects before they occur. Fulfilling the anticipatory function can be difficult when the day-to-day pressures to respond to regulatory deadlines can take most of, if not all, an EPA leader's time and attention. Hence, anticipatory activities will need to be pursued in collaboration with other government agencies, the private sector, and academic engineers and scientists.

Anticipating Environmental and Health Effects of New Technologies

One example of EPA's efforts to identify emerging challenges has been the engagement of its National Advisory Council for Environmental Policy and Technology (NACEPT). NACEPT is an external advisory board established in 1988 to provide independent advice to the agency on a variety of policy, technology, and management issues. The advisory council recently identified several challenges that EPA will need to focus on in the future (EPA NACEPT 2009). The most important challenges identified included climate change, biodiversity losses, and the quality and quantity of water resources. NACEPT also identified corresponding organizational needs for EPA to meet existing and emerging environmental challenges, including improving its ability to use technology more effectively, to transfer technology for commercial uses, and to enhance communication in and outside the agency. The committee concurs with the advisory council's observations that although EPA has demonstrated the ability to create and implement solutions to new challenges in some cases, emerging challenges need to be approached in a more integrated and multidisciplinary way. The committee also concurs with NACEPT's recommendation that EPA include "environmental foresight" or "futures analysis" activities as a regular component of its operations.

Some of EPA programs, including its New Chemicals program and Design for the Environment program (see Chapter 3), already demonstrate strategies for anticipating and mitigating future problems (Tickner et al. 2005). In those programs, EPA has used information on what is known about chemical hazards to develop a series of models so that chemical manufacturers and formulators can predict potential hazards and exposures in the design phase of chemicals. The models are updated as new knowledge emerges. The Design for the Environment example demonstrates that EPA will be best able to address emerging issues through enhanced interdisciplinary collaboration and by using systems thinking and enhanced analysis tools to understand the human health and ecologic implications of important trends. Addressing emerging issues should include consideration of the full life cycle of products, establishment of large-scale surveillance systems to address relevant technologies and indicators, and the analytic ability to detect historical trends rapidly.

Although EPA has engaged NACEPT and its Science Advisory Board (SAB) to help in anticipating trends and has individual programs designed to address concerns about existing and emerging technologies and identify promising new technologies (see, for example, EPA 2011a), the agency does not appear to have a systematic and integrated process for anticipating emerging issues. The example of engineered nanomaterials (discussed below and described in Chapter 3) illustrates some of the problems and pitfalls of current approaches to emerging technologies. A better understanding of such technologies can help to identify and avert ecosystem and health effects and in some cases to avoid unwarranted concern about new technologies that pose little risk.

In principle, early consideration of environmental effects in the design of emerging chemicals, materials, and products offers advantages to businesses, regulatory agencies, and the public, including lower development and compliance costs, opportunities for innovations, and greater protection of public health and the environment. Yet, despite nearly 15 years of investment in engineered nanotechnology and the use of nanomaterials in thousands of products, recognition of potential health and ecosystem effects and design changes that might mitigate the effects have been slow to arrive. Indeed, a December 2011 report by the EPA Office of Inspector General (EPA 2011b) found several limitations in EPA's evaluation and management of engineered nanomaterials and stated the following:

- "Program offices do not have a formal process to coordinate the dissemination and utilization of the potentially mandated information.
- "EPA is not communicating an overall message to external stakeholders regarding policy changes and the risks of nanomaterials.
- "EPA proposes to regulate nanomaterials as chemicals and its success in managing nanomaterials will be linked to the existing limitations of those applicable statutes.
- "EPA's management of nanomaterials is limited by lack of risk information and reliance on industry-submitted data."

The Office of Inspector General concluded that "these issues present significant barriers to effective nanomaterial management when combined with existing resource challenges. If EPA does not improve its internal processes and develop a clear and consistent stakeholder communication process, the Agency will not be able to assure that it is effectively managing nanomaterial risks."

How EPA arrived at that situation provides important information for the design and evaluation of new and emerging technologies. EPA was actively working with other agencies to make large investments in nanotechnology during implementation of the 21st Century Nanotechnology Research and Development Act. In particular, the agency saw the opportunity to use nanotechnology in remediation and funded this type of research. However, it missed the opportunity to support research that addressed proactively the environmental health and safety of nanomaterials, pollution prevention in the production of nanomaterials, and the use of nanotechnology to prevent pollution. In early years, the agency focused primarily on the applications of nanomaterials and not on the environmental and health implications. When it did begin to address implications, the agency focused its attention on defining nanomaterials and whether they are subject to new policy structures because of size-specific hazards (an issue that is still discussed) and on cataloging and redirecting existing research and resources toward assessing exposure, hazard, and risk. The private sector has been left looking for signals from the agency about how it should develop and commercialize nanoscale products.

There were several reasons for the delay in early intervention in the case of nanotechnology. One reason was that materials innovators were focused on discovering new materials and promoting applications of them. Another reason is that materials innovators often have little expertise or formal training in environmental, health, and safety issues. Some of these innovators assumed that nothing about nanomaterials presented new challenges for environmental health and safety and that these were secondary matters to be considered only after commercial products are developed. A third reason was that there was insufficient federal agency leadership, emphasis, and policy regarding *proactive* rather than *reactive* approaches to safer design. Even with increasing knowledge about the design of environmentally benign engineered nanomaterials, progress toward incorporating greater consideration of health and safety in nanomaterial design has been limited for a variety of reasons, including the lack of design rules or other guidance for designers in developing safer technologies, the lack of expertise in solutions-oriented research in EPA, and the lack of collaboration between material innovators and toxicologists and environmental scientists.

The case of engineered nanomaterials indicates the need for EPA to establish more coherent technology-assessment structures to identify early warnings of potential problems associated with a wide range of emerging technologies. If EPA is going to play a major role in promoting and guiding early intervention in the design and production of emerging chemicals (through green chemistry), materials, and products, it will need to commit to this effort beyond its regulatory role.

Many new chemicals and technologies hold considerable potential to improve environmental quality, and it may prove useful for EPA to take some specific steps to anticipate and manage new technologies that emerge from the private sector. Some of these specific steps can be done in collaboration with other agencies, industries, and research organizations when possible. They include:

- Develop baseline design guidelines for new chemicals and technologies and fund research that can anticipate potential effects as part of technology development.
- Balance near-term research that is focused on understanding the potential risks posed by chemicals and technologies that are closer to commercialization with substantial development of longer-term predictive, anticipatory approaches for understanding the potential effects of the technologies.
- Establish processes to collaborate with external partners in academe and industry to attain needed expertise in the development of common metrics for evaluation of emerging technologies.
- Establish opportunities that educate and bring together chemical and materials innovators and environmental health and safety experts (and other stakeholders) to collaborate in understanding and intervening in chemical and materials design.
- Support efforts to amass and disseminate data, models, and design guidelines for safer design to guide emerging technologies.
- Embrace imperfect or incomplete information to guide actions. Uncertainty will always exist in the case of emerging technologies, and identifying alternative paths for action would allow EPA to act or provide guidance for development and commercialization in the face of incomplete data.

Anticipating Emerging Challenges, Scientific Tools, and Scientific Approaches

In recent years, EPA has had to make decisions on several headline-grabbing environmental issues with underdeveloped scientific and technical information or short timelines to gather critical new information, for example, during natural disasters. EPA will always need the capacity to respond quickly to surprises, in part by maintaining a strong cadre of technical staff who are firmly grounded in the fundamentals of their disciplines and able to adapt and respond as new situations arise. But the agency also needs to scan the horizon actively and systematically to enhance its preparedness and to avoid being caught by surprise. Anticipating new scientific tools and approaches will allow EPA to fulfill its mission more effectively.

Collaboration is critical for identifying and addressing many of the topics discussed in Chapter 2, such as trends in energy and climate change and "emerging" environmental concerns that are not new but are the result of improvements in detection capabilities. For example, critics have suggested that the agency's slow response to growing scientific concern about effects of pharmaceutical and personal-care products in surface waters was due in part to its lack of infrastructure or collaboration to address problems that span media and jurisdictions (Daughton 2001). EPA's efforts to anticipate science needs and emerging tools to meet these needs cannot succeed in a vacuum. As it focuses on organizing and catalyzing its internal efforts better, it will need to continue to look outside itself—to other agencies, states, other countries, academe, and the private sector—to identify relevant scientific advances and opportunities where collaboration that relies on others' efforts can be the best (sometimes the only) means of making progress in protecting health and the environment.

Finding: Although EPA has periodically attempted to scan for and anticipate new scientific, technology, and policy developments, these efforts have not been systematic and sustained. The establishment of deliberate and systematic processes for anticipating human health and ecosystem challenges and new scientific and technical opportunities would allow EPA to stay at the leading edge of emerging science.

Recommendation: The committee recommends that EPA engage in a deliberate and systematic "scanning" capability involving staff from ORD, other program offices, and the regions. Such a dedicated and sustained "futures network" (as EPA called groups with a similar function in the past), with time and modest resources, would be able to interact with other federal agencies, academe, and industry to identify emerging issues and bring the newest scientific approaches into EPA.

Science That Innovates

Given EPA's mission and stature as the leading government environmental science and engineering organization, it is imperative that it innovate and support innovation elsewhere in technologies, scientific methods, approaches, tools, and policy instruments. "Innovation" can be challenging to define for a regulatory agency, but one component involves advancing the ability of the agency to discover and characterize problems at a systems level and to provide decision-makers with solutions that are effective and that balance the multiple objectives relevant to the agency and society. Spinoffs from innovation within the agency and activities to promote innovation outside the agency can help environmental authorities in states and other countries to solve their problems and can encourage the regulated community to discover less expensive, faster, and better ways to meet or exceed mandated compliance. Based on the above perspective and using analogies to the typical business definition of innovation, the section below considers processes by which EPA can incorporate and promote innovation.

Identifying Opportunities and Meeting Desired Customer Outcomes

Innovations typically begin with two processes: the identification of opportunity and the understanding of desired "customer" outcomes. An opportunity is simply a "gap" between the current state and a more desirable situation as envisioned by customers. The gaps can be technologic in nature (for example, the need for the design of a new sensor to measure something of interest) or related to a process or business (for example, the need for an approach to obtain up-to-date information from stakeholders). Once an opportunity has been identified and analyzed, an understanding of desired customer outcomes is needed to create innovative solutions.

Understanding desired outcomes goes well beyond simply talking to customers; it includes putting oneself in the clients' shoes to separate what they *say* they want from what they *want*. A common mistake in trying to innovate is to substitute desired *producer* outcomes for desired *customer* outcomes. While EPA is in a different position from product manufacturers, only by understanding why customers are purchasing products can the agency help promote creative solutions. One example is the development of alternative plasticizers for polyvinyl chloride plastics rather than alternative materials that do not require plasticizers. Another example is the creation of less toxic flame retardants rather than creation of an inherently flame-retardant fabric or even consideration of whether flame retardancy is needed for a particular part or product. Insightful, unbiased determination of desired customer outcomes is crucial for proper support of innovation.

An innovative means of defining desired customer outcomes is ethnography, hypothesis-free observation of customers in their "natural habitats". The technique, pioneered by such design firms as IDEO (Palo Alto, CA), has produced a number of insights into consumer behavior that have been translated into successful products. For EPA, the analogue of ethnography is the willingness of staff to visit their "customers" (for example, industry, the general public, or even specific EPA regional offices or laboratories) to see technology or science needs, to see where current regulations or prescribed methods cause people to struggle to conform, or to see where regulations create perverse results. An example of the benefits of observing customer needs is the design of the copying machine. In the 1970s, Xerox used anthropology graduate student Lucy Suchman to observe how users interacted with their copying machines. Suchman created a video showing senior computer scientists at Xerox struggling to make double-sided copies with their own machines. Surprising ethnographic results like that have led to a host of innovative alterations in office equipment that render the user experience much more productive (Suchman 1983). While direct observation of this sort may be unusual for a regulatory agency, similar observational activities by EPA might lead to insights regarding how consumer products are actually used (informing exposure models) or whether responses to specific regulations have unintended consequences that could be readily addressed.

In business, innovation is a catalyst for growth. Business innovation involves the development of ideas or inventions and their translation to the commercial sphere. Innovation results in rapid (favorable) change in market size, market share, sales, or profit through the introduction of new products, processes, or services. Those are clear outcomes that are relatively easy to measure. In an agency like EPA, innovation plays a different role but one that is no less important for the success of the agency in achieving its mission, adapting to changing conditions, and maintaining its authoritative status. Innovation can be thought of as "the conversion of knowledge and ideas into a benefit, which may be for commercial use or for the public good."[1] For the purposes of EPA, the committee is using the term *innovation* as a new means by which to achieve enhancements to environmental and public health at reduced private-sector and public-sector costs. It is essential for EPA to identify and focus on desired outcomes rather than being tied to established processes, procedures, or routines; a fundamental lesson from research on business innovation is that the process is best served by a focus on outcomes.

The simplest measures of success are advances toward goals like cleaner air or safer drinking water, which are most often guided by legislation. Given the scarcity of resources for environmental protection and given the concern for income and employment, EPA has an interest in the private-sector and public-sector costs of achieving health and environmental goals. For EPA, innovation can be measured in such outcomes as direct benefits to health and the environment or in reductions in private-sector and public-sector costs of achieving these outcomes. Continuing to strive to create and promote new processes, tools, and technologies can advance such outcomes. The agency can be innovative in relation to health and the environment by influencing current business and government practices via technology transfer and education.

US Environmental Protection Agency Supporting Innovation

EPA has done much in the past to support the development of innovative ideas in portions of its activities. One example is the development of ways of evaluating and using rapidly emerging biologic testing, as described in Chapter 3. Another is the recent launch of an internal competition called Pathfinder Innovation Projects, which promotes innovation in the agency (EPA 2011c). The program received 117 proposals from almost 300 scientists after its first call for proposals and, after an external peer-review process, funded 12 initial projects (Preuss 2011). Such programs as Design for the Environment, the agency's recent efforts to crowdsource some questions through the Innocentive Web site (described in the section "Identifying New Ways to Collaborate" below), and new technologies in hydroinformatics are examples of efforts to identify innovative solutions. In addition, the federal government's Open Government Initiative and its Challenge.Gov Web site are encouraging innovation in all agencies. Those efforts, however, have not been systematic, and they have not been developed strategically to encourage the much larger potential innovation that could come from the private sector.

When the outcomes are of mutual interest, the agency can help to support and encourage private-sector innovation to serve its desired outcomes in several systematic ways. First, agency scientists generally have a broad view of emerg-

---

[1] http://www.creativeadvantage.com/innovation-definition.aspx

ing toxicologic and human health data—data that may suggest that particular current commercial products are problematic from a health and safety perspective. Highlighting or publicizing such data could provide early hints to manufacturers that replacements may be needed in the future and prompt enterprising companies and entrepreneurs to work to develop alternative means of satisfying desired customer outcomes. The committee does not advise EPA to try to develop solutions, inasmuch as the research environment in the agency is unlikely to be able to duplicate the resources and competitive pressures that drive the commercial product-development market. But providing clear signals of potential future environmental opportunities to the commercial sphere may be enough to prompt the creation of improvements.

Second, EPA can and does provide resources to support private-sector innovation directly. Examples are the EPA Small Business Innovation Research program and enhanced awards programs, such as the Presidential Green Chemistry Awards. Targeting of such programs to address problems that EPA scientists find particularly intractable or to address problems that it does not have the capacity to address can be a valuable means of stimulating the entrepreneurial community to attack problems of direct interest to the agency. To have the resources needed to support private-sector innovation directly at the levels necessary to produce results, the agency would benefit from collaborating and partnering with other agencies that have far greater budgets and resources and similar or complementary innovation challenges, for example, the National Institute of Standards and Technology (NIST), the Department of Energy, and the Department of Defense.

Third, EPA could create an infrastructure that would enable its scientists to serve as a clearinghouse for new technologies, particularly technologies whose effects could cross traditional disciplinary boundaries. The goal of such an infrastructure would be to foster diffusion and adaptation of new technologies, often the slowest step in the innovation process. Steps taken to enhance diffusion could accelerate innovation.

Fourth, technologic innovation relies on willingness (laws and market pressures), capacity, and opportunities for change (Ashford 2000). Capacity becomes a large barrier to innovation adoption, particularly for small and medium-size firms that may not have resources to implement or monitor change or that have legitimate concerns about failed technology adoption. EPA has an important role in addressing capacity and opportunity through science and support that provides information, technical assistance, networking of firms, demonstration activities, and economic incentives and disincentives (Ashford 2000). Many capacity-support mechanisms work most effectively at a state level. Since the passage of the Pollution Prevention Act of 1990, EPA has worked closely with providers of the NIST Manufacturing Extension Partnership and state pollution-prevention technical assistance providers to support innovative adoption of the act. Such models as the Massachusetts Toxics Use Reduction Program provide examples of how an agency like EPA can leverage resources to support innovation. The focus of the program is not on identifying "acceptable" exposure levels but rather options for reducing toxic-substance use in the first place, with science as a driver of innovation.

Literature that discusses and analyzes incentive prize competitions continues to emerge (Kalil 2006; Stine 2009). The federal government is relatively new to this arena and most agencies are still figuring out how to use prizes to fulfill their missions. As EPA is already discovering, using incentives can be a successful way to drive innovation for mission-related topics.

Leveraging Environmental Protection Agency Actions to Promote Private-Sector Innovation

Both intentional and unintentional actions by EPA can affect the willingness of the private sector to invest in research and development. There have not been formal analyses of the extent of such private investment, but it probably dwarfs the investment made by EPA itself. EPA has the potential to expand the investment in new and innovative science and engineering dramatically if it provides signals that are clear, selects instruments and polices that achieve a specific set of outcomes or performances, and allows the regulated community to benefit from innovations (Jaffe et al. 2002, 2005; Popp et al. 2010).

Throughout EPA's history, its actions have resulted in substantial investment in new science and engineering by the private sector, at times with beneficial results. Those actions have taken at least three forms:

- *Regulations.* EPA regulations specify results that need to be achieved and dates by which they need to be achieved. Regulations have, at times, resulted in substantial innovation that might not have been achieved without such clear signals. An example is vehicle carbon monoxide emission standards, which have resulted in substantial investment in developing and continually enhancing the three-way catalyst and dramatically decreasing ambient carbon monoxide despite large increases in travel (NRC 2003).
- *Testing Protocols and Risk Assessment.* In its pursuit of risk estimates for a wide array of substances, EPA can strongly influence research and development investment by the private sector. For example, recent efforts by EPA to enhance its investment in computational toxicology and high-throughput screening have resulted in substantial private investment as well.
- *Public Information.* In requiring the public release of information on emissions and discharges, EPA can set strong incentives for private investment in both major process redesign and product substitution to shift to more sustainable production inputs. One cogent example is the Toxics Release Inventory, which collects data on the disposal and release of over 650 toxic chemicals that are submitted by over 20,000 regulated facilities each year; EPA makes the information available through a publicly accessible Web site. It is an example of consumer-driven change that has led to important reductions in some chemical emissions after its initial public release.

However, the process by which EPA provides incentives for private-sector investment and innovation is not without its challenges. Among them are

- *Overly Prescriptive Rules.* Regulations that use a true performance standard for emissions and discharges can encourage innovation; rules that, in essence, base emission standards on the best current technology (without regular updating) can take away all private incentives for further investment in research and development. For example, the "categorical pretreatment standards" for industrial wastewater discharges locked into place standards based on technologies that were available at the time of promulgation, whereas the "best available control technology" requirements of the Clean Air Act are a "rolling" standard, expressed as performance-based emissions limits that can advance as technology improves. Economic research on innovation and environmental regulation finds that flexible policy instruments that provide rewards for continual environmental improvement and cost reduction tend to promote innovation whereas policies that mandate a specific behavior can deter innovation (Popp et al. 2010).
- *Defensive Rule-Making.* In the current climate in which nearly every action taken by EPA is challenged, the rules that are issued may be written in a conservative fashion that hews tightly to narrow interpretations of the statutes or to past practice and thus may be less likely to encourage innovation once implemented.
- *Reliance Solely on Existing Testing Protocols.* To meet toxicity-testing requirements, EPA often specifies testing protocols in detail, generally on the basis of the state of the art. That practice reduces the incentive to innovate in testing and assessment because of the difficulty of getting new approaches and results accepted.

There are several examples where EPA has been successful in leveraging private sector research. One example is in the Technology and Economic Assessment panels of the Montreal Protocol and the various research and development consortiums designed to find substitute chemicals and technologies for ozone depleting substances (EPA 2007, 2010a). Another example is the Green Lights program for energy efficient light bulbs, the Energy Star program for energy efficient appliances, and the Golden Carrot program for energy efficient refrigerators (EPA 1992; Feist et al. 1994; EPA 2011d; Energy Star 2012). If those examples could be replicated in other situations, EPA would be able to mobilize more industry research and development and implementation to protect the environment.

Finding: EPA has recognized that innovation in environmental science, technology, and regulatory strategies will be essential if it is to continue to perform its mission in a robust and cost-effective manner. However, to date, the agency's approach has been modest in scale and insufficiently systematic.

Recommendation: The committee recommends that EPA develop a more systematic strategy to support innovation in science, technology, and practice.

In accomplishing the recommendation above, the agency would be well-advised to work on identifying more clearly the "signals" that it is or is not sending and to refine them as needed. Clearly identifying signals could be accomplished by seeking to identify the key desired outcomes of EPA's regulatory programs and communicate the desired outcomes clearly to the private and public sector. The committee has identified several ways in which EPA could address this recommendation:

- Establish and periodically update an agency-wide innovation strategy that outlines key desired outcomes, processes for supporting innovation, and opportunities for collaboration. Such a strategy would identify incentives, disincentives, and opportunities in program offices to advance innovation. It would highlight collaborative needs, education, and training for staff to support innovation.
- Identify and implement cross-agency efforts to integrate innovative activities in different parts of the agency to achieve more substantial long-term innovation. One immediate example of such integration that is only beginning to occur is bringing the work on green chemistry from the Design for the Environment program together with the innovative work on high-throughput screening from the ToxCast program to improve application of innovative toxicity-testing tools to the design of green chemicals.
- Explicitly examine the effects of new regulatory and nonregulatory programs on innovation while ascertaining environmental and economic effects. Such an "innovation impact assessment" could, in part, inform the economic evaluation as a structure that encourages technologic innovation that may lead to long-term cost reductions. The assessment could also function as a stand-alone activity to evaluate how regulations could encourage or discourage innovation in a number of activities and sectors. It could help to identify what research and technical support and incentives are necessary to encourage innovation that reduces environmental and health effects while stimulating economic benefits.

Science That Takes the Long View

As the committee has emphasized, the nature and scope of environmental challenges are changing rapidly, as are the scientific and technologic tools and concepts for dealing with them. For instance, the importance of nanoparticles was not evident 2 decades ago. The problems that EPA will face 1 or 2 decades from now are certain to include some challenges that we cannot imagine today. But environmental-protection science in EPA has, for the most part, focused on effects over shorter periods, in single media, or over small spatial scales. That is understandable given regulatory demands for science. However, if EPA is to better understand long-term implications of human effects on ecosystems and health, it will need to develop scientific processes that take the *long view*—that is, processes that can assess changes, even minor ones, over the long term. To detect trends in environmental and human health conditions and to know whether they fall within the range of recent natural variation, long-term data on the basic functioning of environmental systems and human well-being are needed. For example, the scientific community is aware of recent changes in weather patterns, especially increases in extreme events, only because long-term weather records are available.

Indicators

A concise set of environmental indicators can provide information about the status of and trends in key components of natural and human systems and provide evidence of changes that should be monitored. A modest number of environmental indicators of fundamental ecologic processes and attributes are in use (Orians and Policansky 2009; see Box 4-1 for a list of principles to guide the development of indicators). In 2002, a committee of EPA's SAB developed a framework for assessing ecologic conditions (EPA SAB 2002) that is similar to frameworks developed by the H. John Heinz Center (Heinz Center 2002, 2008) and a National Research Council (NRC) committee (NRC 2000). The framework organizes a large number of potential indicators into six categories that represent the key attributes of an ecologic system as a whole. Each attribute can be represented by an individual indicator or by an index created by combining indicators. The six categories can also be used as a checklist for designing environmental management and assessment programs and as a guide for aggregating and organizing information. In its 2008 *Report on the Environment*, EPA analyzed 85 indicators related to environmental and human health that focused on air, water, land, human exposure and health, and ecologic conditions (EPA 2008). However, it has not been clear that the agency is committed to or has a plan to sustain this effort over the longer term. Furthermore, the NRC Committee on Incorporating Sustainability in the US Environmental Protection Agency (NRC 2011a) found that most indicators chosen by EPA are inadequate for exploring the relationship between economic conditions and ecosystem pressure and did not measure such important elements as environmental justice. The committee called for the development of additional sustainability indicators that could include social and economic conditions and, given the challenges of predicting long-term data, stated that any uncertainties in the understanding of indicators should be clearly communicated.

According to some analyses, it is important that EPA continue to develop and adapt a few indicators that are capable of detecting long-term changes in environmental conditions and human well-being above the inevitable noise of variability (GAO 2004). Such indicators should be designed to provide information on basic processes that are most likely to be useful in dealing with both current and future challenges, many of which are unknown today. Indicators whose utility can be evaluated retrospectively would add value in constructing long-term datasets, but EPA should not restrict development of indicators to those with historical data. Many agencies will probably be able to use indicator data, so collaboration among federal agencies, including EPA, would support the collection of data and operationalization of the system of indicators in a more cost-effective manner.

The committee endorses the principles in Box 4-1 to guide the development and use of indicators by EPA and other government agencies that can inform long-term trends. Some of these principles are already used by EPA.

BOX 4-1 Principles to Guide the Development of Indicators

- *Do not ask an indicator to do too much.* Indicators inform us of trends in some entity of interest, but they should not be designed to diagnose the processes that underlie trends.
- *Do not design indicators to give grades.* Indicators should report objective, scientific information, describe trends, and provide the scientific rationale for interpreting them; value judgments should be kept separate from the scientific and objective aspects of indicators.
- *Do not let indicator development be driven by availability of data.* Aware of this trap, the Heinz Center (2002) focused on identifying the environmental processes and products that society most needed to know about by including many empty graphs and explanatory text that directed attention to the processes where ignorance most mattered and where increased research and funding would yield social benefits.
- *Propose and use only a few indicators.* Thousands of environmental indicators have been proposed or are in use. However, an indicator is likely to have the influence it deserves only if it competes for attention with a small number of others.
- *Embed indicators in a rigorous archival system.* Any dataset will be of little value in the absence of a well-crafted system that monitors data quality, document sampling, and analytic methods; archives data in a secure and recoverable form; and analyzes and reports data in formats that are useful to decision-makers and managers.
- *Try to avoid shifting baselines.* Because many ecosystems and habitats are poorly understood, and because large fluctuations characterize most natural environments, choosing appropriate baselines is challenging. For some purposes, a shifting baseline is appropriate, but gradual environmental deterioration is likely to be undetected if a shifting baseline is used, because the altered, often degraded, condition becomes accepted as "normal" (Pauly 1995).
- *Base indicators on well-established scientific principles and concepts.* It is difficult or impossible to interpret the indicator data in the absence of a sound conceptual model of the system to which it is applied.

(Continued)

Building Science for Environmental Protection in the 21st Century 125

> **BOX 4-1 Continued**
>
> - *Develop indicators that are robust and reliable.* A robust indicator is relatively insensitive to expected sources of disturbance and yields reliable and useful numbers in the face of inevitable external perturbations. A robust indicator is based on measurements that can be continued in compatible form when measurement technologies change.
> - *Understand each indicator's statistical properties so that changes in its values will have clear and unambiguous meanings.* The indicator should be sensitive enough to detect real and important changes but not so sensitive that its signals are masked by natural variability.
> - *Clarify the spatial and temporal scales over which each indicator is relevant.* If indicator data is to be aggregated to yield measures on larger spatial scales, consistency in how the data are gathered in different places is vital.
> - *Identify the skills needed at all stages of indicator development and use.* Acceptance of the indicator requires that potential users have confidence in the skills and integrity of the people that gather, store, and report the data.
> - *Separate the entities that compute and report status and trends in indicators from management and enforcement agencies.* Confidence in the numbers being reported requires the belief that the numbers do not depend on who gathers and reports them. Actual or perceived conflicts of interest are likely to arise if the gatherers and interpreters of data also establish and enforce regulations based on trends in the data.
>
> More detail on the reasoning behind the principles above can be found in NRC (2000) and Orians and Policansky (2009).

Long-Term Data Collection

Once indicators have been established, there is a need to measure them over time. To meet its mission, EPA needs an understanding of long-term changes in the environment and trends in rates at which pollutants enter the environment. In the absence of trend and duration data, it is often hard to know whether any specific pollutant load—particularly the load of a nontoxic pollutant, such as nitrogen—is of concern. Long-term monitoring is essential for tracking changes in ecosystems and populations to identify, at the earliest stage, emerging changes and challenges. Without long-term data, it is difficult to know whether current variations fall within the normal range of variation or are truly unprecedented. It is also essential for knowing whether EPA's management interventions are having their intended effect. Monitoring is a fundamental component of hypothesis-testing. All management interventions are based on explicit or implicit hypotheses that justify them and explain why they should yield the desired results. The hypothesis may focus on physical and biologic processes or on expected human behavioral responses. If the hypothesis is made explicit and monitoring is designed specifically to test it, both the value of the monitoring and the details of its design will be clarified and the importance of the monitoring will be evident.

In addition, knowing the pattern of chronic or sporadic exposure of humans and ecosystems to pollutants is essential for understanding their effects. But such an understanding is possible only with the availability of long-term reliable data on pollution loads. Collecting high-quality long-term environmental data on pollutant exposure and ecosystem structure and function is not easy. It might take a decade or more to understand the implications of the trends and the meaning of periodic events. On a practical level, long-term monitoring seldom has general public or political supporters advocating for it, and it is an easy target of budget cutting because it is slow to yield insights.

With the exception of some air and water monitoring programs, there are few long-term monitoring programs, let alone programs that are systematic and rigorous. The paucity of data has made it difficult or impossible to identify key trends related to problems and improvements in environmental quality. That lack of high-quality long-term data is largely the product of four factors:

- Environmental variability across the United States means that what is most useful to monitor differs widely from one place to another.
- It is easy to collect data but much more difficult to collect consistent data, particularly over decades. For example, what is collected may change in response to immediate regulatory needs, thereby reducing its value.
- Over long periods, it is difficult to maintain high-quality data collection systems with solid quality assurance and quality control, well thought-out collection sites, and appropriate collection frequency.
- Monitoring is expensive and often does not produce high-impact information in the short to medium term.

Long-term environmental datasets that have been collected effectively illustrate both the challenges and the rewards of long-term monitoring programs and the importance of collaborations among agencies and organizations. The datasets include those on acid rain, on the Great Lakes ecosystem, and on US Geologic Survey stream-gauging and water-sampling. A key challenge for EPA's science programs is to determine what environmental characteristics to monitor. The answer is tied to indicators, asking the right questions, and ensuring that long-term funding is available to provide the data necessary to support science-based regulatory decisions.

New technologies enable some environmental characteristics to be measured over time across a large spatial domain, for example, satellite imaging and other remote-sensing technologies (as discussed in Chapter 3). The combination of environmental-monitoring data and medical-history information from electronic medical-records data could help to track environmental exposures of human populations and evaluate health effects and dose–response relationships between environmental stressors and health outcomes. The benefits of collaboration, discussed in several places in this report, apply to monitoring.

Finding: It is difficult to understand the overall state of the environment unless one knows what it has been in the past and how it is changing over time. Typically this can only be achieved by examining high-quality time series of key indicators of environmental quality and performance. Currently at EPA, there are few long-term monitoring programs, let alone programs that are systematic and rigorous.

Recommendation: The committee recommends that EPA invest substantial effort to generate broader, deeper, and sustained support for long-term monitoring of key indicators of environmental quality and performance.

Science That Is Collaborative

EPA is a world leader in producing and using science for informed environmental protection, but many other public and private parties in the United States and around the world are also making important contributions in environmental sciences and engineering. Many other parties are working outside the conventional environmental science and engineering space but may have technologies, methods, or data streams that could prove valuable for environmental protection. EPA needs to enhance its ability to draw on those other resources, collaborate with others, and offer leadership, especially in issues that are critical for informing its present missions and for providing the understanding that it will need to address future environmental problems.

Collaborating Among Agencies

Over the years, many investigators in the United States and around the world have looked to EPA to provide leadership in identifying important and emerging problems in environmental science and technology. Individual contact by EPA scientists can help to influence and steer the work of others, but more formal strategies are also needed to influence and direct the focus of research conducted outside EPA. The STAR grants and other extramural support have helped to do that in the context of US universities, but these extramural awards are smaller than those for most research activities relevant to EPA's mission. In some circumstances, EPA may want to consider enhancing its efforts to proactively identify opportunities to collaborate with other federal agencies and national laboratories when practical. In other circumstances, EPA may want to place more focus on clearly articulating the importance of specific research topics to support EPA needs and improve environmental protection more generally.

There has for some time been an established mechanism for coordinating science among agencies of the federal government. It is accomplished through interagency committees of the Office of Science and Technology Policy, such as the Committee on Environment, Natural Resources, and Sustainability (OSTP 2012). In that setting, agencies that are engaged in science that is relevant to EPA's needs come together to exchange information, identify priorities, and plan joint efforts to address key science needs. But the mechanism falls short of what is needed to organize and conduct sustained and successful collaborative efforts, especially in the face of increasing budget constraints and emerging environmental and public health challenges. Furthermore, agencies operate under different, sometimes conflicting statutes, and have varied standards of evidence and scientific needs, which can lead to additional barriers to collaboration. EPA participates in a number of collaborative research efforts with other agencies, such as children's health initiatives with the National Institute of Environmental Health Sciences and the National Nanotechnology Initiative, but future environmental challenges will require much more aggressive efforts to establish and support collaboration. Productive external collaboration should involve a set of proactive steps that include clear mandates from it and other agency leaders and a willingness to understand the regulatory frameworks, strengths, and resource limitations that other agencies face.

Sharing Experiences with Others

EPA maintains world-class laboratories that can serve as a vehicle to induce leading scientists from outside the agency to collaborate with scientists in EPA. EPA also can gain valuable experience and knowledge if its scientists have the ability to work in the research programs and specialized laboratories of other leading research organizations. Such collaborations in either direction can be facilitated through individual arrangements, but it is also important for the agency to continue to support and encourage fellowships that allow outsiders to work with it, university adjunct appointments that allow agency scientists to maintain substantive associations with leading research universities, and a variety of similar programs. That is especially important in addressing problems in which the agency does not have all of the relevant expertise. It will also be important to establish formal mechanisms by which the insights from these collaborations can be shared and infused throughout the agency.

Supporting International Collaboration

As globalization intensifies, domestic action alone will not be enough to address environmental concerns fully, and how other countries protect their environment has an important effect not only domestically but around the world. For example, air pollution, persistent organic pollutants, and mitigating climate change are major challenges that the entire global community faces in the 21st century. They are long-range, transboundary issues; no single nation can solve the problems, and no nation can escape the consequences. Nations have come to recognize that they can protect their own national interests only when the community of nations is able to protect the commons through sustained international collaboration. EPA has identified a variety of objectives for international collaboration, including building strong science institutions, improving access to clean water, and improving urban air quality. The agency works to achieve those objectives by establishing collaborations and partnership with other nations and international organizations. EPA provides resources, tools, and technologies to support international initiatives. Its involvement in international collaboration is not simply one of supporting developing nations but learning from both developed and developing nations about the most innovative technologies and approaches for environmental protection.

For example, EPA is a leading partner in the Partnership for Clean Indoor Air, to which almost 600 partner organizations from over 120 countries are contributing their expertise and resources to reduce exposure to combustion products of fuels used in household cooking and heating (The Partnership for Clean Indoor Air 2012). Indoor smoke from solid fuels poses one of the top 10 health risks globally, contributes an estimated 3.3% of the global disease burden (WHO 2009), and is a source of effects on global climate through emissions of black carbon (Bond et al. 2004; Venkataraman et al. 2005). Since 2002, the partnership has made profound and broad progress in providing clean, efficient, affordable, and safe cooking technologies through commercial markets; reducing indoor air pollution by adopting improved cooking technologies, fuels, and practices; and monitoring and evaluating the health, economic, and environmental effects of the new energy technologies. The Partnership for Clean Indoor Air has also led to better understanding of indoor air pollution due to smoke from burning solid fuels. The mitigation strategies from the partnership have clearly shown both health and environmental benefits (Smith et al. 2009; Wilkinson et al. 2009).

EPA maintains a leadership role in developing science and technology and in translating scientific results to practice and daily life. Maintaining that leadership role can be accomplished by setting priorities for international collaboration with an emphasis on long-range concerns and long-term partnership; establishing multitier collaborations and partnerships with not only foreign governments but industries, academic institutions, and nongovernment organizations in other countries; and maintaining strong leadership in the dissemination of information, the provision of technical expertise, the implementation of policy, and the ability to receive such information globally and to integrate it into practice. As existing challenges persist and new ones emerge, opportunities and challenges for international collaboration will also evolve for EPA. International collaboration is no longer an option; it is a necessity for global solutions to global concerns. International collaboration should be viewed not as a public service or an aid to developing countries but as a crucial mechanism for improving the domestic environment and to gain critical research and implementation skills. It is about maximizing global resources to protect the environment globally and domestically at the same time.

Identifying New Ways to Collaborate

Collaboration within EPA, between agencies or other domestic institutions, and between countries will be increasingly important for addressing the complex problems of the 21st century, but it can be challenging to implement. Incentive structures need to be appropriately aligned, and there need to be mechanisms to facilitate collaboration among individuals or institutions that have different disciplinary backgrounds, are geographically distributed, and have different goals and objectives. Some collaboration will occur within single disciplines, the primary objective being to share knowledge and best practices. Others will seek to exchange knowledge across multiple disciplines, and this may require substantial sustained work.

Regardless of the goal, one way to achieve collaboration is to create "scientific exchange zones" for promoting interaction between disciplines, between scientists and nonscientists, and between strategic research programs (Gorman 2010). Creating such scientific exchange zones involves

- Allowing learning of the languages of multiple disciplines (for example, social science, physical science, water science, risk science, and decision science), which can be done via fellowships, internships, or short-term deployment from one program to another.
- Defining common science questions and establishing common descriptors.
- Creating new and common research methods.
- Identifying those who have top interactional expertise and training the next generation in interactional expertise.
- Developing and supporting experiential interactive projects.

Advances in information technologies (such as those discussed in Chapter 3 and Appendix D) are increasing opportunities for scientific exchange zones. Physicist Michael Nielson identified two ways in which online tools can advance science—by expanding the array of scientific knowledge that can be shared throughout the world and by changing the processes and scale of creative collaboration (Nielsen 2012). Nielson argues for extreme openness in which "as much information as possible is moved out of people's heads and labs, onto the network" where it can be effectively used.

The scientific community has been generally slow to embrace that type of sharing of knowledge, in part because of longstanding views about the need to maintain proprietary methods and databases to enhance the reputation of experts within focused content areas, a key criterion for promotion and tenure. However, federally funded projects increasingly require mechanisms for sharing of methods and databases, and universities and other institutions are developing structures to reward collaborative research. EPA science would benefit from adopting best practices of institutions that are trying to reward collaborative and open-exchange research. There is a growing number of examples of fostering innovation through open communication and collaboration. For example, the Web site InnoCentive is an "open innovation and crowdsourcing pioneer that enables organizations to solve their key problems by connecting them to diverse sources of innovation including employees, customers, [and] partners" (InnoCentive 2012). It uses a "challenge-driven innovation" method that supports innovation programs. Another example of a collaborative-network approach is the National Center for Ecological Analysis and Synthesis, which supports research across disciplines, uses existing data to address ecologic challenges and challenges in allied fields, and encourages the use of science to support management and policy decisions.

Collaboration can also take the form of interaction with members of the general public (which may include people who have scientific expertise). As discussed in Chapter 3, massive online collaboration, also known as crowdsourcing, involves issuing an open call that allows an undefined large group of people or community (crowd) to address a problem or issue that is traditionally addressed by specific individuals. With a well-designed process, crowdsourcing can help to assemble quickly the data, expertise, and resources required to perform a task or solve a problem by allowing people and organizations to collaborate freely and openly across disciplinary and geographic boundaries.

The idea behind regulatory crowdsourcing is that almost every kind of regulation today, from air and water quality to food safety and financial services, could benefit from having a larger crowd of informed people helping to gather, classify, and analyze shared pools of publicly accessible data—data that can be used to educate the public, enhance science, inform public policy-making, or even spur regulatory enforcement actions. Today, a growing number of regulatory agencies (including EPA, the US Securities and Exchange Commission, and the US Food and Drug Administration) see social media and online collaboration as a means of providing richer, more useful, and more interactive pathways for participation. EPA is no stranger to crowdsourcing. Indeed, for the 2009 Toxic Release Inventory, EPA released preliminary data to the public to utilize crowdsourcing as a means for improving and refining the data. The public right-to-know dimension of TRI provided an early example of using informational approaches to encourage environmental change, and also spurred the development of sites like MapEcos.org and Scorecard.org, which provide visual Web-based interfaces that enable citizens to see toxic emissions data and more in one place.

There are several opportunities for crowdsourcing or citizen science (the involvement of the general public in monitoring or other forms of data collection) to augment or enhance EPA scientific and regulatory capabilities, including crowdsourced data collection, urban sensing, and environmental problem-solving. In some domains, EPA would be poised to launch efforts in the near term on the basis of its experiences and existing infrastructure. In others, there would need to be investment in key technologies or resources to make the efforts practical and informative.

Finding: Research on environmental issues is not confined to EPA. In the United States, it is spread across a number of federal agencies, national laboratories, and universities and other public-sector and private-sector facilities. There are also strong programs of environmental research in the public and private sectors in many other nations.

Recommendation: The committee recommends that EPA improve its ability to track systematically, to influence, and in some cases to engage in collaboration with research being done by others in the United States and internationally.

The committee suggests the following mechanisms for approaching the recommendation above:

- Identify knowledge that can inform and support the agency's current regulatory agenda.
- Institute strategies to connect that knowledge to those in the agency who most need it to carry out the agency's mission.
- Inform other federal and nonfederal research programs about the science base that the agency currently needs or believes that it will need to execute its mission.
- Seek early identification of new and emerging environmental problems with which the agency may have to deal.

Crosscutting Example of an Opportunity to Stay at the Leading Edge of Science

As EPA strives to conduct science that anticipates, innovates, takes the long view, and is collaborative, it will be useful for the agency to draw on recent examples to understand in practical terms how it might apply these approaches effectively and in an integrated fashion. The committee describes one such example above in the discussion of the emergence of nanotechnologies and how EPA can better anticipate new technologies. Another broader example, which cuts across all aspects of improving EPA science, is the issue of hydraulic fracturing of shale for natural gas (or hydrofracking). See Box 4-2.

ENHANCED TOOLS AND SKILLS FOR APPLYING SYSTEMS THINKING TO INFORM DECISIONS

Leading-edge science will produce large amounts of new information about the state of human health and ecologic systems and the likely effects of introducing a variety of pollutants or other perturbations into the systems. In particular, many multifactorial problems require systems thinking that can be readily integrated into other analytic approaches (which use risk-assessment concepts for components of the analysis but incorporate other information). Over the years, the agency has become more accomplished in addressing cross-media problems and avoiding "solutions" that transfer a problem from one medium to another, for example, changing an air pollutant to a water or solid-waste pollutant. However, future problems will go beyond cross-media situations and will need to consider global climate and local air quality, land-use patterns and environmental degradation, and implications for industry, the public, and the environment.

BOX 4-2 Putting It All Together: The Case of Hydraulic Fracturing

The set of technologies involved in hydrofracking have implications for many of EPA's programs. The development and operation of hydrofracking facilities can affect surface and ground water, soil, air quality, and greenhouse gas emissions. More broadly, the availability of growing quantities of economically-competitive natural gas can influence industry choices in response to EPA air quality regulations and other rule makings (for example, utility decisions to replace coal-fired electric generating facilities with combined-cycle natural gas in response to EPA emissions rules). Natural gas availability may also have important impacts on other segments of the economy (for example, transportation would be impacted with the development of natural gas infrastructure).

Over the last several years, EPA has become increasingly involved in investigating hydrofracking, both on its own and in concert with a number of federal agencies. It has responded to local issues raised by the activity (often through regional offices), and it has considered and implemented new regulations on the activity (such as, the recent air quality regulations requiring "green completions" for facilities) (Weinhold 2012). However, getting "ahead" of the activities and implementing studies and other actions has been increasingly controversial. For example, in response to FY2010 appropriations language, the agency launched a study of the potential impacts of hydrofracking on groundwater (EPA 2011e), which has been very closely monitored and criticized by industry (Batelle 2012).

The case of hydrofracking gives EPA an opportunity to consider how its science can anticipate, innovate, take the long view, and collaborate, and how it can better embrace systems thinking. It also gives the agency an opportunity to examine how it did or did not apply the concepts presented in the section "Staying at the Leading Edge of Science" and what it might do differently in the future. Such an examination could try to address the questions posed below, among others.

Anticipate: Hydrofracking emerged in the first decade of this century as a rapidly growing means of natural gas (and some oil) production, first in the western United States and in Texas, and then, beginning in 2007, in the

(Continued)

BOX 4-2 Continued

northeast. Its production has grown from a few wells in the beginning to thousands of wells over the last 5 to 10 years. How well did the agency "see" this rapid development coming? Did it hear from its "ears to the ground" in the regional offices and recognize the issue needed an agency-wide approach? How quickly did it grasp both water and air implications? How quickly did it understand the potential need to revisit both its research and regulatory activities?

Innovate: Innovation can be important in something like hydrofracking in a number of ways. For example, assessing complex hydrogeologic systems to understand potential groundwater contamination requires a set of advanced technical skills and familiarity with the latest technologies. At the same time, understanding the potential biologic and ecologic effects of the large number of chemicals being used in hydrofracking requires relatively rapid action, necessitating a decision on the applicability and utility of tools (potentially including life-cycle assessment, health impact assessment, and high-throughput screening) and techniques to evaluate chemical mixtures. How has EPA met these and other needs for innovation in this case? In addition to their own actions, how well have they brought on board the skills and experience of other agencies and the private sector?

Take the long view: While there has been a primary focus on potential shorter-term effects of hydrofracking, it is likely, as with many cases of potential groundwater contamination, that the full potential for contamination can only be determined with a commitment to long-term monitoring around the facilities. EPA has been part of a government-wide effort to coordinate hydrofracking activities (for example, working with the US Geological Survey on long term ground water monitoring). But to what extent is the agency looking at any of its relevant permitting and other authorities and considering how to build long-term monitoring and disclosure into all actions? Such an activity would help to build an essential long-term database.

Is collaborative: There has rarely been an issue that touches on so many public agencies at the federal, state, and local level. The US Centers for Disease Control, National Institutes of Health, US Geological Survey, Department of Energy, state and local environment and health agencies, and many others (including the private sector) are engaged in a wide range of testing, research, and other activities necessary to assess potential risk. How well has the agency applied the principles and ideas described above to enhance its collaboration on an issue like hydrofracking? What could it do to improve that collaboration?

Beyond these four important attributes of leading edge science, hydrofracking also raises a number of broader challenges related to systems thinking that are illustrative of the need for EPA to better embrace such thinking in all it does. For example, to what extent should EPA be stepping back from the near-term water-quality and air-quality issues to ask more fundamental systems questions such as: What are the life-cycle implications of natural

(Continued)

> **BOX 4-2 Continued**
>
> gas for greenhouse gas emissions (such as methane emissions) and how do they compare on a life-cycle basis with other alternatives? From a sustainability point of view, are there ways in which consumers could be encouraged to decrease their consumption of energy that comes from natural gas rather than simply increasing the production of natural gas? Questions such as these are of course beyond the sole domain of EPA, but systems thinking can help inform EPA's scientific research and ultimately its regulatory choices as well.
>
> This case example is not designed to be prescriptive or to suggest that the agency has not been pursuing many of the questions. Rather, a systematic look at the experience with hydrofracking can lend guidance on many fronts for enhancing EPA science's ability to stay at the leading edge and embrace systems thinking in a variety of important fields.

Many analytic tools and skills can contribute to analyzing and evaluating such complex scenarios. The committee describes below four areas in which the agency's tools and skills can be enhanced and integrated to support systems thinking better:

- Life-cycle assessment (LCA).
- Cumulative risk assessment.
- Social, economic, behavioral, and decision sciences.
- Synthesis research.

These tools can be used in conjunction with one another and as inputs to methods for synthesis and evaluation for decisions. In each situation, it is important to integrate efforts to characterize both human health and ecosystem effects.

Life-Cycle Assessment

LCA is "a technique to assess the environmental aspects and potential impacts associated with a product, process, or service, by: compiling an inventory of relevant energy and material inputs and environmental releases; evaluating the potential environmental impacts associated with identified inputs and releases; [and] interpreting the results to help [decision-makers] make a more informed decision" (EPA 2006a, p.2). Performing such analysis requires an accounting of where all materials used in an activity originate and end up. It also requires an accounting of all the inputs into the activity (such as energy and transportation) and their associated environmental consequences and of the changes in other behaviors and other activities that the primary activity induces. Box 4-3 discusses an example of the need for and challenges of LCA.

The idea of LCA is appealing, but the technical details of how to do it well are very challenging. Broadly, two approaches are traditionally used. Process-based LCA is a bottom-up approach that involves itemization of each step in producing a product and consideration of everything from extraction through production and disposal. Although informative and readily interpretable, it systematically underestimates environmental effects by missing key secondary and "ripple" effects (Majeau-Bettez et al. 2011). Data are often inadequate, and strategies to figure out the best way of drawing system boundaries need attention. In addition, although the life-cycle inventory can be constructed in many situations, determining the health or ecologic effects can be challenging given the array of pollutants, the broad scope, and the resulting lack of site specificity of emissions or effects. Researchers have developed approaches to integrating health risk-assessment concepts into process-based LCA, taking account of such factors as pollutant partition coefficients, stack height, and population density to refine the characterization of effects (Humbert et al. 2011), but more work clearly is needed. The second approach involves conducting input–output LCA, in which large matrices of transfers between economic sectors are constructed. That allows consideration of the full ripple effects of actions that are influencing a specific sector (Majeau-Bettez et al. 2011) but with even greater challenges in linking outputs of economic-sector activity with defined health and environmental effects.

EPA has some internal capacity in LCA, has been required to conduct LCA of fuels in the Energy Independence and Security Act of 2007, and has developed tools such as the Tool for Reduction and Assessment of Chemical and Other Environmental Impacts (Bare 2011); but LCA has not been systematically applied to the agency's mission. LCA tools and inventories have been much further developed and applied in other regions, such as Europe (Finnveden et al. 2009). Nonetheless, even without undertaking a formal quantitative LCA, complex systems-level challenges require that the agency at least apply "life-cycle

---

BOX 4-3 The Need for and Challenges of
Life-Cycle Assessment: The Biofuels Case

The need for and challenges of LCA are seen in the case of biofuels. Some analyses suggest that regulatory requirements regarding the use of such fuels may not reduce carbon dioxide emissions and indeed might even increase them (NRC 2010). Those analyses suggest that such mandates could result in a loss of US crop lands available for food production because of the use of the land to produce fuel. That, in turn, could result in pressures to clear forest land in other parts of the world (which is an example of indirect land-use effects) (Searchinger et al. 2008). In addition, the fertilizer to grow such fuel crops in the midwestern United States may contribute to runoff that exacerbates the anoxic zone in the Gulf of Mexico (Rabalais 2010). Thoughtful analysis and interpretation of the results of LCA for biofuels are necessary because some of its methods and assumptions remain controversial (Khosla 2008; Kline and Dale 2008).

thinking" to characterize where a particular product, action, or decision may shift effects somewhere in the life cycle of a product or activity and how those effects can be minimized or prevented. For example, a simple chemical substitution may result in the use of a new product that may be safer for consumers but may cause effects on workers far upstream in the production process. In addition, LCA is an inherently comparative tool because it considers the life-cycle implications of multiple products or processes that achieve the same end use. This so-called *functional* unit determination is intended to be broad and to encourage innovation in the development of solutions by focusing on what a consumer needs from a product rather than on the product itself. Box 4-3 outlines the opportunities that LCA or life-cycle thinking can provide to enhance systems thinking about complex problems.

Cumulative Risk Assessment

The advent of new science tools and techniques means that the suite of traditional tools need to be reviewed and enhanced for 21st century challenges and opportunities. Quantitative risk assessment has been central to many aspects of EPA's mission for decades. The risk-based decision-making framework proposed in *Science and Decisions: Advancing Risk Assessment* (NRC 2009) offers an opportunity, and detailed recommendations, for the agency to revisit and revamp its current practices. In particular, this would encourage linkages between risk assessment and various solutions-oriented approaches. In addition, as discussed in Chapter 3, a host of rapidly evolving health and ecosystem assessment tools (for example, "-omics" and the exposome) can be applied, with appropriate deliberation, to enhance risk assessment further.

Beyond enhancements in traditional single-chemical risk assessment, many of the trends in both science and risk-assessment practice in recent years involve moving from a single-chemical perspective to a multistressor perspective. EPA has grappled with chemical mixtures for some time, and cumulative risk assessment has come to the forefront of the agency's thinking over the last decade, although the agency has rarely used it. Multiple recent NRC committees have addressed cumulative risk assessment extensively (NRC 2008, 2009), and the present committee concurs with the prior recommendations. Moreover, the committee supports the growing emphasis in EPA on this topic (which includes both intramural and extramural research), noting that these efforts have increasingly emphasized community-based participatory approaches, applications in disadvantaged communities, and use of epidemiologic insight. Nonetheless, although much of the emphasis of previous NRC reports has been on cumulative risk assessment for human health effects, it is possible that insights and approaches from ecosystem-based cumulative impact analyses (required under the National Environmental Policy Act [NEPA]) could be adapted to cumulative risk assessment for human health effects.

Cumulative risk assessment contains many subcategories of exposure, health, and ecologic risk analyses, and it is important for EPA to examine its research portfolio in this domain carefully to ensure that it is well aligned with the ultimate decision contexts. With the increased use of LCA or life-cycle thinking, identification of combinations of exposures associated with processes or technologies would be increasingly common, and methods to characterize the ecologic and human health implications of combined exposures would be valuable. There are potentially valuable applications of advanced biosciences for evaluating various chemical mixtures rapidly, but they would not capture psychosocial stressors and other prevalent community-scale factors that are of increasing interest to the agency and various stakeholders (Nweke et al. 2011). New epidemiologic methods or application of epidemiologic insights can start to address those factors, but today they are limited in the number of stressors and locations with adequate exposure data and sample size that they can accommodate. Advancing methods along both fronts, ideally in a coordinated and mutually reinforcing manner, would be the most fruitful approach.

As EPA concentrates increasingly on wicked problems and broad mandates related to sustainability, narrowly focused risk assessments that omit complex interactions will be increasingly uninformative and unsupportive of effective preventive decisions. The broad challenge before the agency will involve developing tools and approaches to characterize cumulative effects in complex systems and harnessing insights from multistressor analyses without paralyzing decisions because of analytic complexities or missing data.

Social, Economic, Behavioral, and Decision Sciences

Systems thinking involves acknowledgment, up front, that environmental conditions are substantially determined by the individual and collective interactions that humans have with environmental processes. As discussed in Chapter 2, the human drivers of environmental change include population growth, settlement patterns, land uses, landscape patterns, the structure of the built environment, consumption patterns, the mix and amounts of energy sources, the spatial structure of production, and a host of other relevant variables. Social, economic, behavioral, and decision sciences show that those drivers are not independent of the natural environments in which effects occur, and that there are feedbacks, positive and negative, between human and environmental systems (Diamond 2005; Ostrom 1990; Taylor 2009). Environmental science and engineering also provide technologies for altering the relationships between humans and the environment and tools for predicting environmental change in response to changes in social and economic systems. That knowledge is all essential and useful for informing environmental decisions and policies; however, additional knowledge, skills, and expertise are needed. To make well-informed policies and decisions that are sustainable, it is essential to integrate theories of, evidence on, and tools for understanding how people respond to changes in the environment, how people respond to interventions that are designed to alter human behavior to achieve desired social and environmental goals, and how specific policies can be implemented within the constraints of legal rights and strongly held, diverse cultural values.

In recognition of that need, it is evident that contributions from the social, economic, behavioral, and decision sciences are crucial for meeting legislative and executive mandates and finding pathways to fulfill EPA's mission sustainably (that is, cost-effectively and equitably and with the greatest prevention effects). Social, economic, behavioral, and decision scientists have the knowledge and expertise to produce analyses that augment traditional health and ecosystem studies to inform policy-makers and stakeholders of the potential economic and social effects of policy decisions. Such analyses have the potential to elucidate the selection of the best solutions not only for the environment but for society as a whole. Spatially explicit assessments of the effects of policies on wages, employment opportunities, and environmental exposures are crucial for understanding the distribution of the benefits and costs of policies and associated community effects by income class, race, and other characteristics relevant to equity and environmental justice (see, for example, Geoghegan and Gray 2005).

Social, economic, behavioral, and decision scientists can help decision-makers to identify unintended environmental or social consequences of public policies such as through the use of predictive economic modeling integrated with environmental modeling. One example is the identification of adverse effects of economically induced land-use changes that resulted from ethanol and renewable-energy policies on nutrient pollution and greenhouse gas emissions (Searchinger et al. 2008; Hellerstein and Malcolm 2011; Secchi et. al. 2011). The effectiveness of environmental policies can be improved if the heterogeneity of humans, the implications of land use, transportation, and other policies affecting the environment, and general equilibrium feedbacks in economic systems are taken into account (Greenstone and Gayer 2007; Kuminoff et al. 2010; Abbott and Klaiber 2011). Providing such information to decision-makers could avoid unintended environmental or social outcomes of regulations and policies. In addition, social, economic, behavioral, and decision scientists have the knowledge and expertise to analyze consumer and business behavior to find less expensive, more effective, and fairer ways to achieve environmental goals (both in the context of existing legislation and in the context of fundamental policy innovations). For example, research with agent-based simulation models (Roth 2002; Duffy 2006; Tesfatsion and Judd 2006; Zhang and Zhang 2007; Parker and Filatova 2008) and laboratory and field experiments (Roth 2002; Suter et al. 2008) are sources of new economic insights for policy instrument design.

For EPA, social, economic, behavioral, and decision science skills can enhance several types of activities that support decisions, including regulatory impact assessments mandated by Executive Order 12866 and others, estimates of economic and social benefits and costs associated with alternative courses of action, and valuation of health benefits and ecosystem services to inform benefit–cost analysis. EPA has made some strides in improving its efforts in this regard, primarily in its application of economic analysis, but the committee notes three important needs for improvement—the need to better integrate social, economic, behavioral, and decision science in decisions; the need for a renewed research effort to update and enhance health and ecosystem valuation and benefits; and the need for substantially improved staff expertise in this field, especially in the social, behavioral, and decision sciences (see the discussion on this topic in the section "Strengthening Science Capacity" in Chapter 5).

Integrating Social, Economic, Behavioral, and Decision Science Skills

Social, economic, behavioral, and decision sciences can serve many functions that are crucial for meeting legislative and executive mandates and for finding pathways to realize EPA's mission cost-effectively and equitably. But even if the gaps are addressed, the benefits of using economics, social, behavioral, and decision sciences in EPA cannot be fully realized unless these areas of expertise are genuinely integrated into EPA decision-making and decision support. The gaps identified by the committee are compounded further by the need for tools to address systems-level impacts—which are often highly uncertain in nature (such as indirect but interconnected impacts of a particular decision or activity)—and solutions that address root causes of problems.

The process of developing a total maximum daily load (TMDL) for the Chesapeake Bay is an example in which EPA conducted high-quality environmental science but did not adequately integrate social, economic, behavioral, and decision sciences. The TMDL calls for reductions in nitrogen (by 25%), phosphorus (by 24%), and sediment (by 20%) to restore the bay by 2025 and allocates load reductions in its major tributaries to the bay (EPA 2010b). The TMDL can be viewed as a triumph of EPA-led environmental science. The agency initiated and led research to understand the effects of human activity on the bay's waters and living resources and to provide a scientific foundation for measures to restore the bay beginning in the 1970s. That research has been crucial for the development of the science that underpins the TMDL, but the TMDL was developed without studies of the benefits and costs. EPA's National Center for Environmental Economics and its Chesapeake Bay program are only now conducting benefit–cost assessments of the TMDL, which are too late to inform its specification. Furthermore, and perhaps even more problematic, EPA has neither conducted nor sponsored substantial social, economic, behavioral, and decision science research on fundamental policy questions related to inducing the behavioral changes that are essential for achieving the TMDL.

Updating and Enhancing Estimates of Environmental Benefits

Among the social, economic, behavioral, and decision sciences, only economics is generally mandated in EPA. Regulatory impact assessments to determine the benefits and costs of environmental regulation are mandated by various executive orders. The most important is Executive Order 12866, which requires benefit–cost analyses of proposed and final regulations that qualify as "significant" regulatory actions. The Safe Drinking Water Act, the Toxic Substances Control Act, and the Federal Insecticide, Fungicide, and Rodenticide Act require EPA to weigh benefits and costs in regulatory actions. Some environmental legislation requires benefit and cost evaluations outside the regulatory process. The leading example is Section 812 of the Clean Air Act Amendments of 1990, which requires EPA to develop periodic reports to Congress that estimate the economic benefits and costs of provisions of the act; program offices are responsible for regulatory impact assessments in their fields. EPA's National Center for Environmental Economics offers a centralized source of technical expertise for economic assessments in the agency.

Evaluations of EPA economic assessments indicate that they can be useful and influential. For example, an early evaluation of economic assessments (EPA 1987) found that "economic analyses improve environmental regulation. EPA's benefit–cost analyses have resulted in several cases of increased net societal benefits of environmental regulations." The report also found that "benefit–cost analysis often provides the basis of stricter environmental regulations....For example, the most dramatic increase in net benefits ($6.7 billion) from EPA's [regulatory impact assessments] resulted from a recommendation for much stricter standards—to eliminate lead in motor fuels." The report also noted that, "alternatively, benefit–cost analysis may reveal regulatory alternatives that achieve the desired degree of environmental benefits at a lower cost."

There are many uncertain and potentially controversial dimensions associated with the use of benefit–cost analysis as conducted for regulatory impact assessments. In principle, such analyses identify, quantify, and monetize the multiple outcomes of an environmental decision or policy into a single indicator of economic efficiency. If multiple alternatives are considered in the analyses, benefit-cost analyses can support a solutions orientation by incorporating economics factors into the risk-based decision-making paradigm described earlier. Apart from procedural details, there is debate about the validity of economic concepts of value for environmental and some other goods (for example, the value of life), the capacity of economics to measure some types of values, the discounting of future costs and benefits, the treatment of uncertainty and irreversibility, and the relevance of economic efficiency, as one among many societal objectives, to environmental decisions (EPA 1987; Ackerman and Heinzerling 2004; Posner 2004; Sunstein 2005). Despite the controversies, the importance of benefit–cost analysis for regulatory impact assessments is recognized almost universally. Harrington et al. (2009) have produced a useful set of recommendations to improve the technical quality, relevance to decision-making, and transparency of regulatory impact assessments and their treatment of new scientific information and balance of efficiency and distributional concerns. If implemented, a number of those recommendations would help integrate benefit–cost analysis with other tools to support systems thinking, including a focus on comparing multiple policy alternatives, making decisions given multiple dimensions of interest, and improving how uncertainties are characterized and communicated. The issues of multidimensional decision-making and addressing uncertainty in complex systems are discussed below. EPA's economists are cognizant of the controversies and challenges in conducting benefit-cost studies and of the frontiers of economic research in environmental benefit–cost analysis.

Even if benefit–cost analysis were implemented based on the recommendations from Harrington et al. (2009), there are important gaps in the scope of available work on the valuation of benefits, and the literature is becoming dated. For example, a value-of-a-statistical life (VSL) approach is used to assign monetary values to reductions in mortality risk. EPA typically bases its VSL values on a 1992 synthesis of 26 published studies (Viscusi 1992). Although EPA does provide more recent references to frame the discussion, including studies of how VSL may vary as a function of life expectancy or health status, the core quantitative value remains based on old studies that are not necessarily relevant to the people most vulnerable to air-pollution health effects. Inasmuch as analyses have consistently shown that uncertainty in VSL dominates the overall uncertainty in benefit–cost analyses and given that policy choices may hinge on this value, it seems incumbent on EPA to invest in intramural and extramural research specifically on it. Similarly, with respect to morbidity outcomes, the most recent willingness-to-pay study that was incorporated into the analysis of the Clean Air Act Amendments (EPA 2011f) was conducted in 1994. In that benefit-cost analysis, multiple key health outcomes were valued by using only cost-of-illness information.

Valuation of the ecologic and welfare benefits of air-pollution reductions is similarly lacking; the only dimensions monetized are the effects of reductions in agricultural and forest productivity on the price of related goods, the willingness to pay for visibility improvements (based on studies conducted 20–30 years ago), damage to building materials, and effects on recreational fishing and timber in the Adirondacks. A recent workshop on the use of ecologic nonmarket valuation in EPA benefit–cost analysis work concluded that "perhaps the most surprising outcome was the realization of how few nonmarket ecological valuation studies are used by the EPA" (Weber 2010).

Funding for valuation research has been reduced, and disciplinary interest in valuation research, once a major topic in environmental-economics journals, has diminished. Assessing and addressing gaps in the environmental-benefits estimates should have high priority and can be tackled through research designs that produce statistically representative samples for EPA regulatory impact assessments (for the importance of standardization and sampling strategies for water see, for example, Bruins and Heberling 2004; Van Houtven et al. 2007; Weber 2010). The challenges in addressing these gaps are not trivial given budget constraints and logistics barriers to collecting public data.

Two recent EPA documents discuss ecologic-valuation challenges and strategies for the agency (EPA 2006b; EPA SAB 2009). The stated goal of EPA's *Ecological Benefits Assessment Strategic Plan* is to "help improve

Agency decision-making by enhancing EPA's ability to identify, quantify, and value the ecological benefits of existing and proposed policies" (EPA 2006b, p. XV). The agency has devoted resources to enhancing the science of ecologic-service valuation through the STAR grants program and ORD's ecosystem-services research program. The 2009 report by EPA SAB concluded that a "gap exists between the need to understand and protect ecologic systems and services and EPA's ability to address this need" (EPA SAB 2009, p.8). The report provides recommendations for enhanced research on "how an integrated and expanded approach to ecologic valuation can help the agency describe and measure the value of protecting ecologic systems and services, thus better meeting its overall mission" (EPA SAB 2009, p.8).

Synthesis Research

Scientific progress has always depended on synthesis of disparate data, concepts, and theories (Carpenter et al. 2009). The combined forces of increasing research specialization, an explosion of scientific information, and growing demand for solutions to pressing environmental problems have made scientific synthesis more challenging and more urgent than ever before. In recent years, the National Science Foundation and other agencies have invested considerable funds in synthesis research centers. At least 19 such centers have now been established in the United States and abroad. They have demonstrated the power and cost effectiveness of bringing together multidisciplinary collaborative groups to integrate and analyze data to generate new scientific knowledge that has increased generality, parsimony, applicability, and empirical soundness (Hampton and Parker 2011). The impact of well-designed synthesis efforts extends beyond the life of the projects themselves. Projects spin off new and unexpected collaborative research, and researchers tend to expand the multidisciplinary breadth of their research (Hampton and Parker 2011). Several mechanisms that increase the creative productivity of multidisciplinary synthesis research have been identified, notably open, competitive calls for projects; face-to-face interactions at a neutral facility free of distractions; and multiple working group meetings that enable technology and analytic support, institutional diversity, diversity of career stages, inclusion of postdoctoral fellows, and moderately large group size (Hackett et al. 2008; Hampton and Parker 2011).

EPA often produces useful synthesis reports that summarize the state of knowledge on a topic, but this is not a substitute for synthesis *research*. The agency could make more use of deliberately designed synthesis research activities to promote multidisciplinary collaborations and accelerate progress toward integrated sustainability science. One example is the recent creation by the US Geological Survey of the John Wesley Powell Center for Analysis and Synthesis (The Powell Center 2012). EPA could also pursue opportunities with synthesis centers, such as the National Center for Ecological Analysis and Synthesis (NCEAS 2012) and the newly established Socio-Environmental Synthesis Center (SESYNC 2012). Given its corpus of researchers in both environmental and health sciences, the agency is well positioned to pursue synthesis research that brings together environmental science and public-health science data and perspectives.

SYNTHESIS AND EVALUATION FOR DECISIONS

Systems-level problems are rarely amenable to simple quantitative decision measures. More often than not, complex problems require consideration of multiple types of information (including quantitative and qualitative data), characterization of different types of uncertainty, and consideration of prevention options. The information base might include outputs from tools such as LCA or cumulative risk assessment, integrated with economic and other information in a structured framework to inform decisions. There is a need for the agency to develop consistent approaches for synthesizing a broad array of systems information on hazards, exposures, solutions, and values. Although agencies like EPA regularly "do synthesis" for decision-making, the approaches to synthesis have been varied, often depending on regulatory demands. Most recently, EPA has attempted to realign its existing science decision-making processes in line with the sustainability framework proposed by the NRC Committee on Incorporating Sustainability in the US Environmental Protection Agency (NRC 2011a), although implementation of that realignment is in its early stages. The committee identified several approaches that could provide support to the agency in establishing consistent approaches for more holistic decisions. They include enhanced sustainability analysis (as recommended by NRC 2011a), solutions-oriented approaches (such as alternatives assessment and health impact assessment), and multicriteria decision analysis.

Sustainability Analysis

EPA has recently begun to implement tools and approaches to determine how the science that it is developing and decisions made on the basis of it support sustainability (Anastas 2012). The NRC Committee on Incorporating Sustainability in the US Environmental Protection Agency developed a sustainability-analysis framework for EPA (NRC 2011a), starting with the definition of sustainability espoused in Executive Order 13514 (2009). The definition of sustainability provided in that executive order is "to create and maintain conditions, under which man and nature can exist in productive harmony, and fulfill the social, economic, and other requirements of present and future generations of Americans" (42 U.S.C. §§ 4331(a)[NEPA§101]). That committee developed its sustainability framework and the sustainability assessment and management approach (Figure 4-2) to provide guidance to EPA on incorporating sustainability into decision-making. They build on the traditional risk-assessment and risk-management framework of the agency.

The framework and assessment and management approach are built on traditional principles of vision, objectives, goals, and metrics. The goals of sustainability analysis are to expand decision consideration to include multiple sustainability options and their social, environmental, and economic consequences; to include the intergenerational effects of consequences in addition to more immediate ones; and to involve a broad array of stakeholders. Many of these concepts intersect with the solutions-oriented approaches discussed in this section, including the expansive scope and stakeholder involvement that will be discussed in the health impact assessment (HIA) paragraph below, the use of behavioral science and economics to consider an array of impacts, and the use of life-cycle thinking to avoid creating upstream and downstream problems. The framework and approach lay out a series of steps that should be taken in evaluating sustainability implications of a particular decision. The evaluation tools to be used will depend on the nature and needs of the particular decision. Although this framework is new and does not have a particular "toolbox" or analytic technique, it provides a set of steps that can be taken in synthesizing information from varied sources and fields into a coherent sustainability decision.

Solutions-Oriented Approaches

There has been an increasing emphasis among advisory committees and in EPA on moving away from characterizing problems and toward determining and evaluating solutions. For example, *Science and Decisions: Advancing Risk Assessment* (NRC 2009) emphasized that risk assessment should be used to discriminate among risk-management options, not as an end in itself, and this suggests a framework within which alternative options are considered upfront. A recent NRC report (NRC 2011b) gave recommendations about HIA as a solutions-oriented policy tool to introduce health considerations into numerous policy decisions that could have direct or indirect health implications. HIA, as defined by the NRC Committee on Health Impact Assessment (2011b), is consistent with the risk-based decision-making framework proposed by *Science and Decisions: Advancing Risk Assessment* (NRC 2009). Both approaches explicitly emphasize conducting analyses that help discriminate among policy options and that use planning and scoping to devise analyses that are of an appropriate level of sophistication given the decision context. Although it includes approaches beyond risk assessment and has a scope that often extends beyond EPA's mandate, HIA has many attributes that are well-aligned with the future needs of EPA. For example, HIA incorporates systems thinking and encourages development of broad conceptual models to avoid unanticipated risk tradeoffs, which is a valuable approach to incorporate into numerous analytic tools. HIA also endorses the use of both quantitative and qualitative information to inform decisions, and it explicitly considers equity issues and vulnerable populations that may not be captured within benefit-cost analyses or related tools.

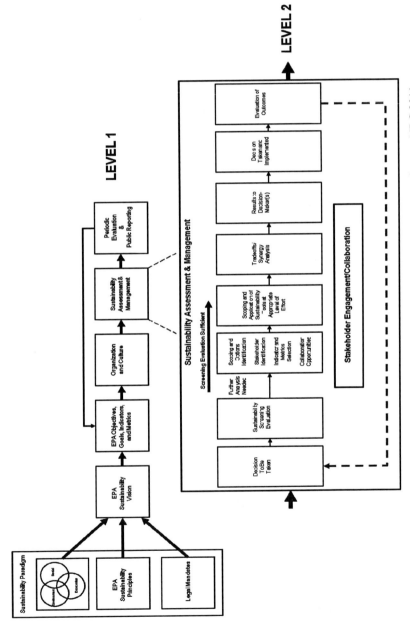

FIGURE 4-2 A framework for sustainable decisions at the US Environmental Protection Agency. Source: NRC 2011a.

In parallel, alternatives assessment has formed the basis of pollution-prevention planning efforts, the chemical-alternatives assessment processes undertaken by the EPA Design for the Environment program (see Chapter 3), and technology options analysis in chemical safety efforts. Although alternatives assessment is not strictly tied to risk assessment and risk management, it similarly involves the systematic analysis of a wide array of options for a potentially damaging activity that are evaluated on the basis of hazard, performance, social, and economic factors. Beyond HIA and alternatives assessment, there are several other tools for applying systems thinking that are intrinsically solutions-oriented. For example, LCA emphasizes comparing alternative methods for addressing a defined need, and benefit–cost analysis is designed to compare multiple policy options to arrive at an optimal choice.

Regardless of the specific approach and application, those approaches all provide a tool for focusing on solutions and innovation opportunities and drawing attention to what a government agency or proponent of an activity could be doing to solve the problem at hand rather than simply characterizing it in finer detail. They also provide opportunities to evaluate the reduction of multiple risks rather than simply focusing on controlling a single hazard, potentially leveraging the methods and approaches within cumulative risk assessment. Finally, if agencies' actions promote restriction of a particular activity, there is a responsibility to understand alternatives and support a path that is environmentally sound, technically feasible, and economically viable and that does not create new risks of its own. Box 4-4 gives an example of a solutions-oriented approach for reducing chemical use.

Many of the above solutions-oriented approaches are currently in use in some manner in EPA, but they are not applied comprehensively and systematically across the agency. However, alternatives-assessment approaches are built into numerous laws and international treaties. The process for carrying out an environmental impact statement under NEPA and state programs is one of the most comprehensive examples for the requirement of alternatives assessment at the national level (Tickner and Geiser 2004). When assessments are undertaken under NEPA, agencies and organizations that use public funds and that are carrying out activities that might have substantial effects on the environment need to undergo the process for creating an environmental impact statement. "The goal of NEPA is to foster better decisions and 'excellent action' through the identification of reasonable alternatives that will avoid or minimize adverse impacts" (Tickner and Geiser 2004).

NEPA regulations require that the process described above be carried out before the start of any activity that might have environmental effects. An interdisciplinary approach is undertaken to ensure that environmental effects and values are comprehensively identified and examined; to ensure that appropriate and reasonable alternatives are rigorously studied, developed, and described; and to recommend specific courses of action. The first step of assessing effects is a scoping process, during which potential effects are broadly defined and examined in detail, "including direct and indirect impacts, cumulative effects, effects on historic and cultural resources, impacts of alternatives, and options to mitigate potential impacts" (Tickner and Geiser 2004). The NEPA environmental impact statement approach, supplemented by new approaches to health impact assessment, provides a way of integrating scientific information from multiple sources into decisions that focus on evaluating prevention options.

BOX 4-4 Example of a Solutions-Oriented Approach: Reducing Trichloroethylene Use in Massachusetts

The solvent trichloroethylene (TCE) has been targeted for substantial reductions in exposure by EPA and numerous states because of its toxicity, particularly its potential carcinogenicity. It is commonly found at Superfund sites and is of particular concern because it can leach into and contaminate groundwater and drinking water supplies. TCE is mainly used to degrease metal parts and it can cause harmful occupational exposures if it is accidentally spilled. Applying traditional end-of-pipe control approaches have, in many cases, resulted in the TCE problem being shifted from air to water to land rather than the problem being eliminated. Reducing human exposures to TCE cannot be solved using a simple solution; a systems-based and solutions-oriented approach must be used.

Under the 1989 Massachusetts Toxics Use Reduction Act, chemical manufacturers that produce large quantities of toxic chemicals, which include TCE, are required to pay a fee and to conduct a form of systems analysis. The analysis includes a materials throughput analysis every year and a facility planning process analysis every 2 years to understand how and why chemicals are being used and to assess potential process and product modifications that would reduce toxic material use and waste. The fee provides funding for the Toxics Use Reduction Institute (TURI) at the University of Massachusetts, Lowell. Most cleaning tasks that use TCE can be performed with alternative organic solvents or with water-based cleaners. In general, water-based cleaners are preferred because they are usually safer for human health and the environment. TURI has been working with manufacturers of metal parts and manufacturers of electronics to help them move from TCE to safer and more cost-effective cleaning solutions. TURI determined that one of the barriers in the adoption of safer alternatives is the concern that productivity and product specifications might suffer if standard metal cleaning procedures are altered. To address this concern, TURI created the Surface Solutions Laboratory to evaluate the effectiveness and safety of TCE alternatives for small-sized and medium-sized companies. By focusing on the "function" that TCE provides, requiring a systems evaluation, and providing support for solutions, industrial TCE use in Massachusetts declined by more than 77% from 1990 to 2005, with greater than 90% reductions in some sectors.

Source: Adapted from Sarewitz et al. 2010 and TURI 2011.

EPA has substantially contributed to the advancement of analytic techniques and tools to detect environmental stressors and characterize health and ecosystem impacts of those stressors. While better characterization of problems is important, it is critical that the agency apply this knowledge to primary prevention—that is, the design of safer and more sustainable forms of production and consumption. Like sustainability, a focus on solutions should be more than a simple mission statement. It must be linked to adequate resources, tools, and infrastructure at the highest levels of the agency.

Multiple-Criteria and Multidimensional Decision-Making

The tools of alternatives assessment, HIA, and the sustainability management approach all incorporate an array of information to arrive at a preferred solution, but this becomes increasingly challenging given numerous dimensions that often cannot be compared on the same scale. Benefit–cost analysis is a well-known example in which the multiple outcomes of a decision are monetized (if possible) and aggregated into a single indicator of economic efficiency, but it cannot provide a complete ranking of alternatives if stakeholders and environmental decision-makers are interested in other objectives (such as fairness across income classes, regions, or racial groups; generations in the distribution of burdens and benefits; or norms in the treatment of nonhuman organisms). Benefit–cost analysis is useful and sometimes mandated for regulatory impact assessments, but its value is limited in dealing with complex issues in which economic efficiency is only one of many important objectives for environmental decision-makers and their stakeholders. While deliberative approaches may be warranted in complex situations, especially when both quantitative and qualitative information are being used, analytic approaches to integrate data from multiple sources and types into a single number or range of numbers have tremendous potential.

One approach to solving problems that have multiple incommensurate dimensions is to use tools within the realm of multiple-criteria decision-making (MCDM) (Figueira et al 2005). Within the broad framework of informatics, developing and applying MCDM in conjunction with uncertainty analysis and data-mining (Shi et al. 2002) can provide a set of useful ways for using emerging science and developing evidence-supported policy-making in the agency. Like benefit–cost analysis, MCDM is an approach that creates and assigns a preference index to rank policy options on the basis of the totality of all adopted criteria. However, unlike benefit–cost analysis, MCDM was not designed to rank options based on a consumer's preference for environmental or other goods. Instead, the method is flexible for selecting weights and it is often designed to use weights assigned by the decision-maker. This flexibility allows for the inclusion of a broader set of objectives, although the selection can be inherently contentious. The preference index value attributable to each criterion reflects the nature and importance of the criterion, for example, cost, benefits, innovation, or change in ecology or human health. The preference index then leads to a partial ranking of the policy options under consideration and recommendation of an "optimal" set of choices or competitive choices (Brans and Vincke 1985). MCDM has been applied successfully in environmental decision-making (Moffett and Sarkar 2006; Hajkowicz and Collins 2007); however, criterion-specific constituents of the preference index for each policy option are affected by the quality of the science and evidence, scaling, and other factors that can limit validity (Hajkowicz and Collins 2007).

An alternative to single-objective formulations is to provide decision-makers with the Pareto optimal set of nondominated candidate solutions. Essentially, the Pareto optimal set is constructed by identifying decisions that can improve one or more objectives without harming any other. Use of the Pareto optimal set does not determine a single preferred approach but presents decision-makers with a smaller set of options from which to choose. The concept of Pareto optimal sets is not new, but the capacity to apply it in decision-making has been greatly expanded by recent methodologic advances in optimization techniques (most notably multiobjective evolutionary algorithms) and computation of Pareto sets for large complex problems, and this has increased the scope of environmental and other applications (Coello et al. 2007; Nicklow et al. 2010). Rabotyagov et al. (2010) give an example of evolutionary computation for the analysis of tradeoffs between pollution-control costs and nutrient-pollution reductions. Optimal sets of air pollution control measures have been developed that consider aggregate health benefits and inequality in the distribution of those benefits as separate dimensions (Levy et al. 2007). Kasprzyk et al. (2009) demonstrate how multiobjective methods can be used to inform policies for the management of urban water-supply risks that are caused by growing population demands and droughts. Multiobjective optimization in support of environmental-management decisions is especially compelling given the emerging paradigm of managing for multiple ecosystem services and consideration of cumulative risks for human health. Tradeoffs and complementarities can exist between alternative services and between other relevant performance metrics (for example, public and private costs and distribution outcomes by location or income class). Applications of multiobjective optimization methods would promote the explicit specification of preference indices relevant to environmental decision-making and science to quantify outcomes and evaluate tradeoffs; all this would serve to improve the transparency and scientific soundness of decisions.

Addressing Uncertainty in Complex Systems

With any of the solutions-oriented approaches delineated above, regardless of which analytic tools or indicators are used by EPA to support decisions in the future, uncertainty will be an overriding concern. With increasingly complex multifactorial problems and a push for tools that are sufficiently timely and flexible to inform risk-management decisions (NRC 2009), the importance of uncertainty characterization and analysis will only increase. It should be noted that the increasing importance of uncertainty analysis does not necessarily imply increasing sophistication of computational methods or even increasing necessity of quantitative uncertainty analysis. As discussed in *Science and Decisions: Advancing Risk Assessment* (NRC 2009), uncertainty analysis is a component to be planned for with the rest of an assessment, and a simple bounding analysis or qualitative elucidation of different types of uncertainties may be adequate if it shows that a given risk-management decision is robust compared with competing options (NRC 2009).

Consistent and holistic approaches are necessary for characterizing and recognizing uncertainty (in particular the various types of uncertainty, including unquantifiable systems-level uncertainties, indeterminacy, and ignorance). Such approaches would allow EPA to articulate the importance of uncertainty in light of pending decisions and not become paralyzed by the need for increasingly complex computational analysis. In addition, applying uncertainty analysis coherently in all EPA's arenas would ensure that a policy or decision is both tenable and robust (van der Sluijs et al. 2008) and would ensure that uncertainty analysis is a means to an end and is designed with the end use in mind. Similarly, uncertainty analyses that are billed as comprehensive but omit key sources of uncertainty have the potential to be misleading or to lead to inappropriate decisions about research priorities and interventions. Finally, EPA would benefit from communicating uncertainty more effectively. Uncertainty is often mistakenly viewed as a negative form of knowledge, an indicator of poor-quality science (Funtowicz and Ravetz 1992). There is therefore a perception that acknowledging uncertainty can weaken agency authority by creating an image of the agency as unknowledgeable, by threatening the objectivity of "science-based" standards, and by making it more difficult to defend itself in the face of political and court challenges. However, reluctance to acknowledge uncertainty can lead EPA to rely on tools and methods that cannot provide timely answers, can push the agency to use point estimates to defend what are policy decisions (see Brickman et al. 1985), and runs counter to the value of uncertainty analysis in informing research and decision priorities.

OVERARCHING RECOMMENDATION

The committee has described the important emerging environmental issues and complex challenges in Chapter 2 and the many types of emerging scientific information, tools, techniques, and technologies in Chapter 3 and Appendixes C and D. It is clear that if EPA is to meet those challenges and to make the greatest possible use of the new scientific tools, its problems will need to be approached from a systems perspective. Although improved science is important for EPA's future, it is not sufficient for fully improving EPA's capabilities for dealing with health and environmental challenges. Better economic analysis, policy approaches, stakeholder involvement, communication, policy, and integration for systems thinking are also vital.

In the present chapter, the committee has recommended ways in which the agency can integrate systems thinking techniques into a 21st century framework for science to inform decisions. For EPA to stay at the leading edge, it will need to produce science that is anticipatory, innovative, long-term, and collaborative; to evaluate and apply new tools for data acquisition, modeling, and knowledge development; to continue to develop and apply new systems-level tools and expertise; and to develop tools and methods to synthesize science, characterize uncertainties, and integrate, track, and assess the outcomes of actions. If effectively implemented, such a framework would help to break the silos of the agency and promote collaboration among research related to different media, time scales, and disciplines. In supporting environmental science and engineering for the 21st century, EPA will need to continue to evolve from an agency that focuses on using science to characterize risks so that it can respond to problems to an agency that applies science holistically to characterize both problems and solutions at the earliest point possible.

Finding: Environmental problems are increasingly interconnected. EPA can no longer address just one environmental hazard at a time without considering how that problem interacts with, is influenced by, and influences other aspects of the environment.

Recommendation: The committee recommends that EPA substantially enhance the integration of systems thinking into its work and enhance its capacity to apply systems thinking to all aspects of how it approaches complex decisions.

The following paragraphs provide examples of some of strategies that EPA could use to help it set its own priorities and to enhance its use of systems thinking.

Even if formal quantitative LCA is not feasible, increased use of a life-cycle perspective would help EPA to assess activities, regulatory strategies, and associated environmental consequences. Placing more of a focus on life-cycle thinking would likely include increasing EPA's investment in the development of LCA tools that reflect the most recent knowledge in LCA and risk assessment (both human health and ecologic). In addition, it may be more cost effective for EPA to provide incentives and resources to increase collaborations between LCA practitioners in the agency and those working on related analytic tools (such as risk assessment, exposure modeling, alternatives assessment, and green chemistry). EPA has some internal capacity for LCA, but could benefit from a more systematic use of such an assessment across the agency's mission.

Continuing to invest intramural and extramural resources in cumulative risk assessment and the underlying multistressor data, including coordinated bench science and community-based components, would give EPA a broader and more comprehensive understanding of the complex interactions between chemicals, humans, and the environment. A challenge before the agency is the characterization of cumulative effects using complex, incomplete, or missing data. Even as EPA seeks to improve its understanding of risks, some prevention-based decisions may need to be made in the face of uncertainty.

In EPA's science programs, environmental decisions will only be effective if they consider the social and behavioral contexts in which they will play out. Such decisions can substantially affect societal interests beyond those that are specifically environmental. Tradeoffs among environmental and other societal outcomes need to be anticipated and made explicit if decision-making is to be fully informed and transparent. Predicting economic and societal responses at various points in the decision-making process is necessary to achieve desirable environmental and societal outcomes. For these reasons, developing mechanisms to integrate social, economic, behavioral, and decision sciences would lead to more comprehensive environmental-management decisions. EPA can engage the social, economic, behavioral, and decision sciences as part of a systems-thinking perspective rather than as consumers and evaluators of others' science. Human behavior is a major determinant of the state of the environment and, as such, should be an integral part of systems thinking regarding environmental risk and risk mitigation alternatives. In addition, EPA would benefit from a long-term commitment to advancing research in a number of related fields, including valuation of health and ecosystem benefits.

Research centers that focus on synthesis research have demonstrated the power and cost effectiveness of bringing together multidisciplinary collaborative groups to integrate and analyze data to generate new scientific knowledge. Deliberately introducing synthesis research into EPA's activities would contribute to accelerating its progress in sustainability science. A specific area where knowledge from systems thinking could be applied is in the design of safe chemicals, products, and materials.

REFERENCES

Abbott, J.K., and H.A. Klaiber. 2011. An embarrassment of riches: Confronting omitted variable bias and multi-scale capitalization in hedonic price models. Rev. Econ. Stat. 93(4):1331-1342.

Ackerman, F., and L. Heinzerling. 2004. Priceless: On Knowing the Price of Everything and the Value of Nothing. New York: The New Press.

Anastas, P. 2012. Fundamental changes to EPA's research enterprise: The path forward. Environ. Sci. Technol. 46(2):580-586.

Ashford, N.A. 2000. An Innovation-Based Strategy for a Sustainable Environment. Pp. 67-107 in Innovation-Oriented Environmental Regulation: Theoretical Approach and Empirical Analysis, J. Hemmelskamp, K. Rennings, and F. Leone, eds. Heidelberg: Springer Verlag [online]. Available: http://18.7.29.232/bitstream/handle/1721.1/1590/Potsdam.pdf?sequence=1 [accessed Apr. 17, 2012].

Bare, J.C. 2011. TRACI 2.0: The tool for the reduction and assessment of chemical and other environmental impacts 2.0. Clean Technol. Environ. Policy 13(5):687-696.

Batelle. 2012. Review of EPA Hydraulic Fracturing Study Plan, EPA/600/R11/122, November 2011. Prepared for American Petroleum Institute and America's Natural Gas Alliance, Washington, DC, by Batelle, Lexington, MA. June 2012 [online]. Available: http://anga.us/media/251570/final_epa_study_plan_review_061112.pdf [accessed July 23, 2012].

Bond, T.C., D.G. Streets, K.F. Yarber, S.M. Nelson, J.H. Woo, and Z. Klimont. 2004. A technology-based global inventory of black and organic carbon emissions from combustion. J. Geophys. Res. 109:D14203, doi:10.1029/2003JD003697.

Brans, J.P., and P.H. Vincke. 1985. A preference ranking organization method: (The PROMETHEE method for multiple criteria decision-making). Manage. Sci. 31(6):647-656.

Brickman, R., S. Jasanoff, and T. Ilgen. 1985. Controlling Chemicals: The Politics of Regulation in Europe and the United States. Ithaca: Cornell University Press.

Bruins, R.J.F., and M.T. Heberling. 2004. Ecological and economic analysis for water quality standards. Pp. 128-142 in Economics and Ecological Risk Assessment: Applications to Watershed Management, R.J.F. Bruins, and M.T. Heberling, eds. Boca Raton: CRC Press.

Carpenter, S.R., E.V. Armbrust, P.W. Arzberger, F.S. Chapin, J.J. Elser, E.J. Hackett, A.R. Ives, P.M. Kareiva, M.A. Leibold, P. Lundberg, M. Mangel, N. Merchant, W.W. Murdoch, M.A. Palmer, D.P.C. Peters, S.T.A. Pickett, K.K. Smith, D.H. Wall, and A.S. Zimmerman. 2009. Accelerate synthesis in ecology and environmental sciences. Bioscience 59(8):699-701.

Coello, C.A., G.B. Lamont, and D.A. van Veldhuizen, eds. 2007. Evolutionary Algorithms for Solving Multi-Objective Problems (Genetic Algorithms and Evolutionary Computation), 2nd Ed. New York: Springer.

Daughton, C.G. 2001. Pharmaceuticals in the environment: Overarching issues and overview. Pp. 2-38 in Pharmaceuticals and Personal Care Products in the Environment: Scientific and Regulatory Issues, C.G. Daughton, and T. Jones-Lepp, eds. Symposium Series 791. Washington, DC: American Chemical Society.

Diamond, J. 2005. Collapse: How Societies Choose to Fail or Succeed. New York: Penguin.

Duffy, J. 2006. Agent-based models and human subject experiments. Pp. 949-1011 in Handbook of Computational Economics, Vol. 2, L. Tesfatsion, and K.L. Judd, eds. Amsterdam: Elsevier [online]. Available: http://www.pitt.edu/~jduffy/papers/duffy 2006.pdf [accessed Apr. 11, 2012].

Energy Star. 2012. About Energy Star [online]. Available: http://www.energystar.gov/index.cfm?c=about.ab_index [accessed July 25, 2012].

EPA (US Environmental Protection Agency). 1987. EPA's Use of Benefit-Cost Analysis 1981-1986. EPA-230-05-87-028. Office of Policy Planning and Evaluation, US Environmental Protection Agency, Washington DC [online]. Available: http://yosemite.epa.gov/ee/epa/eerm.nsf/vwAN/EE-0222-1.pdf/$file/EE-0222-1.pdf [accessed Apr. 11, 2012].

EPA (US Environmental Protection Agency). 1992. The Climate is Right for Action, Voluntary Programs to Reduce Greenhouse Gas Emissions. EPA400K92005. Office of Air and Radiation, US Environmental Protection Agency, Cincinnati, OH. October 1992.

EPA (US Environmental Protection Agency). 2006a. Life Cycle Assessment: Principles and Practice. EPA/600/R-06/060. National Risk Management Research Laboratory, Office of Research and Development, US Environmental Protection Agency, Cincinnati, OH. May 2006 [online]. Available: http://www.epa.gov/nrmrl/std/lca/

pdfs/chapter1_frontmatter_lca101.pdf http://www.epa.gov/nrmrl/std/lca/pdfs/600r06 060.pdf [accessed Aug. 3, 2012].

EPA (US Environmental Protection Agency). 2006b. Ecological Benefits Assessment Strategic Plan. EPA-240-R-06-001. Office of the Administrator, US Environmental Protection Agency, Washington, DC. October 2006 [online]. Available: http://yosemite.epa.gov/ee/epa/eerm.nsf/vwAN/EE-0485-01.pdf/$File/EE-0485-01.pdf [accessed Apr. 12, 2012].

EPA (US Environmental Protection Agency). 2007. Achievements in Stratospheric Ozone Protection. Progress Report. EPA-430-R-07-001. Office of Air and Radiation, US Environmental Protection Agency, Washington, DC. April 2007 [online]. Available: http://www.epa.gov/ozone/downloads/spd-annual-report_final.pdf [accessed July 25, 2012].

EPA (US Environmental Protection Agency). 2008. EPA's Report on the Environment. EPA/600/R-07/045F. National Center for Environmental Assessment, US Environmental Protection Agency, Washington, DC [online]. Available: http://www.epa.gov/roe/docs/roe_final/EPAROE_FINAL_2008.PDF [accessed Apr. 17, 2012].

EPA (US Environmental Protection Agency). 2010a. The Montreal Protocol on Substances that Deplete the Ozone Layer. Ozone Layer Protection, US Environmental Protection Agency [online]. Available: http://www.epa.gov/ozone/intpol/ [accessed July 25, 2012].

EPA (US Environmental Protection Agency). 2010b. Chesapeake Bay Total Maximum Daily Load for Nitrogen, Phosphorus, and Sediment. Chesapeake Bay Program Office, Region 3, Annapolis, MD; Office of Regional Counsel, Region 3, Philadelphia, PA, Division of Environmental Planning and Protection, Region 2, New York, and Office of Water, Office of Air and Radiation, Office of General Counsel, Office of the Administrator, US Environmental Protection Agency, Washington, DC [online]. Available: http://www.epa.gov/reg3wapd/pdf/pdf_chesbay/Final BayTMDL/CBayFinalTMDLExecSumSection1through3_final.pdf [accessed Apr. 16, 2012].

EPA (US Environmental Protection Agency). 2011a. Technology Innovation and Field Services Division. Office of Superfund Remediation and Technology Innovation, US Environmental Protection Agency [online]. Available: http://www.epa.gov/superfund/partners/osrti/tifsd.htm [accessed Jan. 15, 2012].

EPA (US Environmental Protection Agency). 2011b. EPA Needs to Manage Nanomaterial Risks More Effectively. Report No. 12-P-0162. Office of the Inspector General, US Environmental Protection Agency, Washington, DC. December 29, 2011 [online]. Available: http://www.epa.gov/oig/reports/2012/20121229-12-P-0162.pdf [accessed Apr. 10, 2012].

EPA (US Environmental Protection Agency). 2011c. The Pathfinder Projects. US Environmental Protection Agency [online]. Available: www.epa.gov/ord/sciencematters/june2011/pathfinder.htm [accessed Jan. 15, 2012].

EPA (US Environmental Protection Agency). 2011d. Green Lights. Voluntary Programs, Region 7, US Environmental Protection Agency [online]. Available: http://www.epa.gov/region7/p2/volprog/grnlight.htm [accessed July 25, 2012].

EPA (US Environmental Protection Agency). 2011e. Plan to Study the Potential Impacts of Hydraulic Fracturing on Drinking Water Resources. EPA/600/R-11/122. Office of Research and Development, US Environmental Protection Agency. November 2011 [online]. Available: http://water.epa.gov/type/groundwater/uic/class2/hydraulicfracturing/upload/hf_study_plan_110211_final_508.pdf [accessed July 23, 2012].

EPA (US Environmental Protection Agency). 2011f. The Benefits and Costs of the Clean Air Act from 1990 to 2020. Office of Air and Radiation, US Environmental Protection Agency, Washington, DC. March 2011 [online]. Available: http://www.epa.gov/oar/sect812/feb11/fullreport.pdf [accessed Apr. 12, 2012].

EPA (US Environmental Protection Agency). 2012. Science and Technology. US Environmental Protection Agency [online]. Available: http://www.epa.gov/gateway/science/ [accessed Mar. 26, 2012].

EPA NACEPT (US Environmental Protection Agency National Advisory Council for Environmental Policy and Technology). 2009. Outlook for EPA. EPA 130-R-09-002. Office of Cooperative Environment Management, US Environmental Protection Agency. March 16, 2009 [online]. Available: http://www.epa.gov/ofacmo/nacept/reports/pdf/2009_0316_nacept_letter_and_outlook_report.pdf [accessed Apr. 12, 2012].

EPA SAB (US Environmental Protection Agency Science Advisory Board). 2002. A Framework for Assessing and Reporting on Ecological Condition: An SAB Report. EPA-SAB-EPEC-02-009. Science Advisory Board, US Environmental Protection Agency, Washington, DC. June 2002 [online]. Available: http://yosemite.epa.gov/sab/sabproduct.nsf/7700D7673673CE83852570CA0075458A/$File/epec02009.pdf [accessed Apr. 17, 2012].

EPA SAB (US Environmental Protection Agency Science Advisory Board). 2009. Valuing the Protection of Ecological Systems and Services. EPA-SAB-09-012. Science Advisory Board, US Environmental Protection Agency, Washington, DC. May 2012 [online]. Available: http://yosemite.epa.gov/sab/sabproduct.nsf/WebBoard/F3DB1F5C6EF90EE1852575C500589157/$File/EPA-SAB-09-012-unsigned.pdf [accessed Apr. 10, 2012].

Feist, J.W., R. Farhang, J. Erickson, E. Stergakos, P. Brodie, and P. Liepe. 1994. Super Efficient Refrigerators: The Goldon Carrot from Concept to Reality. Paper 3.67 in Proceedings ACEEE Summer Study on Energy Efficiency in Buildings, August 1, 1994. Washington, DC: American Council for Energy-Efficient Economy [online]. Available: http://eec.ucdavis.edu/ACEEE/1994-96/1994/VOL03/067.PDF [accessed Aug. 2, 2012].

Figueira, J., S. Greco, and M. Ehrgott, eds. 2005. Multiple Criteria Decision Analysis: State of the Art Surveys. International Series in Operations Research and Management Science Volume 78. New York: Springer.

Finnveden, G., M.A. Hauschild, T. Ekvall, J. Guinée, R. Heijungs, S. Hellweg, A. Koehler, D. Pennington, and S. Suh. 2009. Recent developments in life cycle assessment. J. Environ. Manage. 91(1):1-21.

Funtowicz, S., and J. Ravetz. 1992. Three types of risk assessment and the emergence of post-normal science. Pp. 251-274 in Social Theories of Risk, S. Krimsky, and D. Golding, eds. Westport, CT: Praeger.

GAO (US General Accountability Office). 2004. Environmental Indicators. Better Coordination is Needed to Develop Environmental Indicator Sets that Inform Decisions. GAO-05-52 [online]. Available: http://www.gao.gov/new.items/d0552.pdf [accessed Aug. 13, 2012].

Geoghegan, J., and W. Gray. 2005. Spatial environmental policy. Pp. 52-96 in International Yearbook of Environmental and Natural Resource Economics 2005/2006: A Survey of Current Issues, H. Folmer, and T. Tietenberg, eds. Northampton, MA: Edward Elgar Publishing.

Gorman, M.E. ed. 2010. Trading Zones and Interactional Expertise: Creating New Kind of Collaboration. Cambridge, MA: MIT Press.

Greenstone, M., and T. Gayer. 2007. Quasi-Experimental and Experimental Approaches to Environmental Economics. Discussion Paper 07-22. Resources for the Future, Washington, DC. June 2007 [online]. Available: http://www.rff.org/documents/rff-dp-07-22.pdf [accessed Apr. 11, 2012].

Hackett, E.J., J.N. Parker, D. Conz, D. Rhoten, and A. Parker. 2008. Ecology transformed: The National Center for Ecological Analysis and Synthesis and the changing patterns of ecological research. Pp. 277-296 in Scientific Collaboration on the Internet, G.M. Olson, A. Zimmerman, and N. Bos, eds. Boston, MA: MIT Press.

Hajkowicz, S., and K. Collins. 2007. A review of multiple criteria analysis for water resource planning and management. Water Resour. Manage. 21(9):1553-1566.

Hampton, S.E., and J.N. Parker. 2011. Collaboration and productivity in scientific synthesis. BioScience 61(11):900-910.

Harrington, W., L. Heinzerling, and R. Morgenstern, eds. 2009. Reforming Regulatory Impact Analysis. Washington, DC: Resources for the Future [online]. Available: http://www.rff.org/RFF/Documents/RFF.RIA.V4.low_res.pdf pdf [accessed Apr. 11, 2012].

Heinz Center (The H. John Heinz III Center for Science, Economics and the Environment). 2002. The State of the Nation's Ecosystems: Measuring the Lands, Waters, and Living Resources of the United States. Cambridge, UK: Cambridge University Press.

Heinz Center (The H. John Heinz III Center for Science, Economics and the Environment). 2008. The State of the Nation's Ecosystems 2008: Measuring the Lands, Waters, and Living Resources of the United States. Washington, DC: Island Press.

Hellerstein, D., and S. Malcolm. 2011. The Influence of Rising Commodity Prices on the Conservation Reserve Program. Economic Research Report No. 110, US Department of Agriculture, Economic Research Services [online]. Available: http://www.ers.usda.gov/Publications/ERR110/ERR110.pdf [accessed Apr. 9, 2012].

Humbert, S., J.D. Marshall, S. Shaked, J.V. Spadaro, Y. Nishioka, P. Preiss, T.E. McKone, A. Horvath, and O. Jolliet. 2011. Intake fraction for particulate matter: Recommendations for Life Cycle Impact Assessment. Environ. Sci. Technol. 45(11):4808-4816.

InnoCentive. 2012. What We Do [online]. Available: http://www.innocentive.com/about-innocentive [accessed Apr. 16, 2012].

Jaffe, A.B., R.G. Newell, and R.N. Stavins. 2002. Environmental policy and technological change. Environ. Resour. Econ. 22(1-2):41-69.

Jaffe, A.B., R.G. Newell, and R.N. Stavins. 2005. A tale of two market failures: Technology and environmental policy. Ecol. Econ. 54(2-3):164-174.

Kalil, T. 2006. Prizes for Technological Innovation. Washington, DC: The Brookings Institution [online]. Available: http://www.brookings.edu/views/papers/200612kalil.pdf [accessed Aug. 2, 2012].

Kasprzyk, J.R., P.M. Reed, B.R. Kirsch, and G. W. Characklis.2009. Managing population and drought risks using many-objective water portfolio planning under uncertainty, Water Resour. Res. 45: W12401, doi:10.1029/2009WR008121.

Khosla, V. 2008. Biofuels: Clarifying assumptions. Science 322(5900):371-374.

Kline, K.L., and V.H. Dale. 2008. Biofuels: Effects on land and fire. Science 321 (5886):199-201.

Kuminoff, N., V.K. Smith, and C. Timmins. 2010. The New Economics of Equilibrium Sorting and its Transformational Role for Policy Evaluation. NBER Working Paper No. 16349. National Bureau of Economic Research, Cambridge, MA [online]. Available: http://www.nber.org/papers/w16349.pdf [accessed Apr. 2, 2012].

Levy, J.I., A.M. Wilson, and L.M. Zwack. 2007. Quantifying the efficiency and equity implications of power plant air pollution control strategies in the United States. Environ. Health Perspect. 115(5):740-750.

Majeau-Bettez, G., A.H. Strømman, and E.G. Hertwich. 2011. Evaluation of process- and input/output-based life cycle inventory data with regard to truncation and aggregation issues. Environ. Sci. Technol. 45(23):10170-10177.

Meadows, D. 1999. Leverage Points: Places to Intervene in a System. Hartland, VT: The Sustainability Institute [online]. Available: http://www.sustainer.org/pubs/Leverage_Points.pdf [accessed Apr. 16, 2012].

Moffett, A., and S. Sarkar. 2006. Incorporating multiple criteria into the design of conservation area networks: A minireview with recommendations. Diversity Distrib. 12(2):125-137.

NCEAS (National Center for Ecological Analysis and Synthesis). 2012. Welcome to NCEAS [online]. Available: http://www.nceas.ucsb.edu/ [accessed Apr. 16, 2012].

Nicklow, J., P. Reed, D. Savic, T. Dessalegne, L. Harrell, A. Chan-Hilton, M. Karamouz, B. Minsker, A. Ostfeld, A. Singh, and E. Zechman. 2010. State of the art for genetic algorithms and beyond in water resources planning and management. J. Water Resour. Plan. Manage. 136(4):412-432.

Nielsen, M. 2012. Reinventing Discovery: The New Era of Networked Science. Princeton, NJ: Princeton University Press.

NRC (National Research Council). 2000. Ecological Indicators for the Nation. Washington, DC: National Academy Press.

NRC (National Research Council). 2003. The Measure of STAR: Review of the US Environmental Protection Agency's Science to Achieve Results (STAR) Research Grants Program. Washington, DC: National Academies Press.

NRC (National Research Council). 2008. Phthalates and Cumulative Risk Assessment: The Task Ahead. Washington, DC: National Academies Press.

NRC (National Research Council). 2009. Science and Decisions: Advancing Risk Assessment. Washington, DC: National Academies Press.

NRC (National Research Council). 2010. Hidden Costs of Energy: Unpriced Consequences of Energy Production and Use. Washington, DC: National Academies Press.

NRC (National Research Council). 2011a. Sustainability and the US EPA. Washington, DC: National Academies Press.

NRC (National Research Council). 2011b. Improving Health in the United States: The Role of Health Impact Assessment. Washington, DC: National Academies Press.

Nweke, O.C., D. Payne-Sturges, L. Garcia, C. Lee, H. Zenick, P. Grevatt, W.H. Sanders, III, H. Case, and I. Dankwa-Mullan. 2011. Symposium on Integrating the Science of Environmental Justice into Decision-Making at the Environmental Protection Agency: An overview. Am. J. Pub. Health 101(S1):S19-S26.

Orians, G.H., and D. Policansky. 2009. Scientific bases of macroenvironmental indicators. Annu. Rev. Environ. Resour. 34:375-404.

OSTP (Office of Science and Technology Policy). 2012. NSTC Committee on Environment, Natural Resources, and Sustainability. Office of Science and Technology Policy [online]. Available: http://www.whitehouse.gov/administration/eop/ostp/nstc/committees/cenrs [accessed Mar. 26, 2012].

Ostrom, E. 1990. Governing the Commons: The Evolution of Institution for Collective Action. Cambridge: Cambridge University Press.

Parker, D.C., and T. Filatova. 2008. A theoretical design for a bilateral agent-based land market with heterogeneous economic agents. Comput. Environ. Urban Syst. 32(6):454-463.
Pauly, D. 1995. Anecdotes and the shifting base-line syndrome of fisheries. Trends Ecol. Evol. 10(10):430.
Popp, D., R. Newell, and A. Jaffe. 2010. Energy, the environment, and technical change. Pp. 873-937 in Handbook of the Economics of Innovation, Vol. 2, B.H. Hall, and N. Rosenberg, eds. Amsterdam: Elsevier.
Posner, R.A. 2004. Catastrophe: Risk and Response. New York: Oxford University Press.
Preuss, P. 2011. EPA Research: A Year of Innovation. Science Matter Newsletter, December 20, 2011. http://www.epa.gov/sciencematters/december2011/executivemessage.htm [accessed Apr. 13, 2012].
Rabalais, N.N. 2010. Eutrophication of estuarine and coastal ecosystems. Pp. 115-135 in Environmental Microbiology, 2nd Ed., R. Mitchell, and J.D. Gu, eds. Hoboken, NJ: John Wiley & Sons.
Rabotyagov, S., T. Campbell, M. Jha, P.W. Gassman, J. Arnold, L. Kurkalova, S. Secchi, H. Feng, and C.L. Kling. 2010. Least-cost control of agricultural nutrient contributions to the Gulf of Mexico hypoxic zone. Ecol. Appl. 20(6):1542-1555.
Roth, A.E. 2002. The economist as engineer: Game theory, experimentation, and computation as tools for design economics. Econometrica 70(4):1341-1378.
Sarewitz, D., D. Kriebel, R. Clapp, C. Crumbley, P. Hoppin, M. Jacobs, and J. Tickner. 2010. The Sustainable Solutions Agenda. The Consortium for Science, Policy and Outcomes, Arizona State University, Lowell Center for Sustainable Production, University of Massachusetts Lowell [online]. Available: http://www.sustainableproduction.org/downlods/SSABooklet.pdf [accessed Apr. 10, 2012].
Searchinger, T., R. Heimlich, R.A. Houghton, F. Dong, F. Elobeid, J. Fabiosa, S. Tokgoz, D. Hayes, and T.H. Yu. 2008. Use of US croplands for biofuels increases greenhouse gases through emissions from land-use change. Science 319(5867):1238-1240.
Secchi, S., P.W. Glassman, M. Jha, L. Kurkalova, and C.L. Kling. 2011. Potential water quality changes due to corn expansion in the Upper Mississippi River Basin. Ecol. Appl. 21(4):1068-1084.
SESYNC (Socio-Environmental Synthesis Center). 2012. Socio-Environmental Synthesis Center [online]. Available: http://www.sesync.org/ [accessed Apr. 16, 2012].
Shi, T., Y. Peng, W. Xu, and X. Tang. 2002. Data mining via multiple criteria linear programming: Applications in credit card portfolio management. IJITDM 1(1):131-151.
Smith, K.R., M. Jerrett, H.R. Anderson, R.T. Burnett, V. Stone, R. Derwent, R.W. Atkinson, A. Cohen, S.B. Shonkoff, D. Krewski, C.A. Pope, III, M.J. Thun, and G. Thurston. 2009. Public health benefits of strategies to reduce greenhouse-gas emissions: Health implications of short-lived greenhouse pollutants. Lancet 374(9707): 2091-2103.
Stine, D. 2009. Federally Funded Innovation Inducement Prizes. CRS Report for Congress, June 29, 2009 [online]. Available: http://www.fas.org/sgp/crs/misc/R40677.pdf [accessed July 20, 2012].
Suchman, L. 1983. When User Hits Machine. Keynote address to CHI'83 Conference on Human Factors in Computing Systems, December 12-15, 1983, Boston.
Sunstein, C.R. 2005. Cost benefit analysis and the environment. Ethics 115(2):351-385.

Suter, J.F., C.A. Vossler, G.L. Poe, and K. Segerson, 2008. Experiments on damage-based ambient taxes for nonpoint source polluters. Am. J. Agric. Econ. 90(1):86-102.

Taylor, M.S. 2009. Environmental crises: Past, present, and future. Can. J. Econ. 42(4): 1240-1275.

Tesfatsion, L., and K.L. Judd, eds. 2006. Handbook of Computational Economics, Vol. 2. Agent-Based Computational Economics. Amsterdam: Elsevier.

The Partnership for Clean Indoor Air. 2012. Welcome. The Partnership for Clean Indoor Air [online]. Available: http://www.pciaonline.org/ [accessed May 3, 2012].

The Powell Center (John Wesley Powell Center for Analysis and Synthesis). 2012. Welcome to the Powell Center [online]. Available: http://powellcenter.usgs.gov/ [accessed Apr. 16, 2012].

Tickner, J.A., and K. Geiser. 2004. The precautionary principle stimulus for solutions- and alternatives-based environmental policy. Environ. Impact Assess. Rev. 24(7-8):801-824.

Tickner, J., K. Geiser, and M. Coffin. 2005. The US experience in promoting sustainable chemistry. Environ. Sci. Pollut. Res. 12(2):115-123.

TURI (Toxics Use Reduction Institute). 2011. Use Nationally and in Massachusetts [online]. Available: http://www.turi.org/About/Library/TURI_Publications/Massach usetts_Chemical_Fact_Sheets/Trichloroethylene_TCE_Fact_Sheet/TCE_Facts/Use_ Nationally_and_in_Massachusetts [accessed July 24, 2012].

van der Sluijs, J.P., A.C. Petersen, P.H.M. Janssen, J.S. Risbey, and J.R. Ravetz. 2008. Exploring the quality of evidence for complex and contested policy decisions. Environ. Res. Lett. 3(2):024008.

Van Houtven, G., J. Powers, and S.K. Pattanayak. 2007. Valuing water quality improvements in the United States using meta-analysis: Is the glass half-full or half-empty for national policy analysis? Resour. Energy Econ. 29(3):206-228.

Venkataraman, C., G. Habib, A. Eiguren-Fernandez, A.H. Miguel, and S.K. Friedlander. 2005. Residential biofuels in South Asia: Carbonaceous aerosol emissions and climate impacts. Science 307(5714):1454-1456.

Viscusi, W.K. 1992. Fatal Tradeoffs: Public and Private Responsibilities for Risk. New York: Oxford University Press.

Weber, M. 2010. EPA Use of Ecological Nonmarket Valuation. AERE Newsletter 30(1):26-35 [online]. Available: http://www.aere.org/newsletters/documents/May 2010Newsletter.pdf [accessed Apr. 13, 2012].

Weinhold, B. 2012. The Future of fracking: New rules target air emissions for cleaner natural gas production. Environ. Health Perspect. 120(7):A272-A279.

WHO (World Health Organization). 2009. Global Health Risks: Mortality and Burden of Disease Attributable to Selected Major Risks. Geneva: WHO Press [online]. Available: http://www.who.int/healthinfo/global_burden_disease/GlobalHealthRis ks_report_full.pdf [accessed Aug. 3, 2012].

Wilkinson, P., K.R Smith, M. Davies, H. Adair, B.G Armstrong, M. Barrett, N. Bruce, A. Haines, I. Hamilton, T. Oreszczyn, I. Ridley, C. Tonne, and Z. Chalabi. 2009. Public health benefits of strategies to reduce greenhouse-gas emissions: Household energy. Lancet 374(9705):1917-1929.

Zhang, T., and D. Zhang. 2007. Agent-based simulation of consumer purchase decision-making and the decoy effect. J. Bus. Res. 60(8):912-922.

5

Enhanced Scientific Leadership and Capacity in the US Environmental Protection Agency

Previous chapters, particularly Chapter 4, outline the need for an enhanced approach to science and technology in the US Environmental Protection Agency (EPA) that recognizes the challenge of characterizing and preventing effects on human health and ecosystems in the context of complex systems. With the development of new tools and approaches to collecting and processing large amounts of environmental and health data and for characterizing effects when knowledge is uncertain, it is imperative that a new way of thinking—embodied in the concepts of science that anticipates, innovates, takes the long view, and is collaborative—be integrated into scientific processes in EPA's Office of Research and Development (ORD) and across its national research program areas.

In the United States, environmental management is conducted through a mosaic of federal, state, and local activities in multiple federal and state agencies, often through regionally distributed offices. Environmental decisions are made at multiple administrative levels in those agencies. Science questions arise throughout that environmental-management network and require access to the latest and best scientific information possible. In EPA's program and regional offices, science is most often conducted in direct response to particular regulatory and programmatic needs and often operates on different timescales in contrast with longer-term discovery-oriented science in ORD. Efforts to enhance EPA science for the 21st century should not focus only on ORD but should incorporate efforts, resources, expertise, and scientific and non-scientific perspectives in program and field offices. Such efforts need to support the integration of both existing and new science throughout the agency; avoid duplication or, worse, contradictory actions; respect different sets of priorities and timeframes; and advance common goals. EPA also engages in activities to deliver science and provide decision support to nonfederal entities (for example, states and tribes), and decreasing budgets of tribal, state, and local environmental agencies will make this function increasingly important. At the same time, EPA is itself increasingly resource constrained. As noted in *Science Integration for Decision Making at the US Environmental Protection Agency*, since 2004, the budget for ORD has declined 28.5% in real dollar terms (gross domestic product-indexed dollars) (EPA SAB 2012a).

To support enhanced leadership and to continually improve environmental science and engineering for the 21st century, the committee identified six key topics:

- Enhance agency-wide science leadership.
- Fully implement the recent restructuring of ORD.
- Coordinate and integrate science efforts within the agency more effectively.
- Strengthen scientific capacity inside and outside the agency.
- Deliver and support 21st century environmental science and engineering outside the agency.
- Support scientific integrity and quality.

ENHANCED AGENCY-WIDE SCIENCE LEADERSHIP IN THE US ENVIRONMENTAL PROTECTION AGENCY

Emerging challenges in ecosystem quality and human health necessitate the enhancement and broader use of science in the agency. The environmental challenges outlined in Chapter 2, such as climate change and degradation of surface waters from mixtures of contaminants, share many characteristics—they are transboundary, are multigenerational, and involve complex interactions of multiple stressors and feedback loops. They are affected by population growth, changes in land-use patterns, and technologic change. They constitute wicked problems—that is, problems that are difficult to characterize and to solve because of their complexity; lack of comprehensive understanding; controversy over causes, effects, and solutions; and interdependence. The rapidly emerging scientific techniques and approaches and their application described in Chapter 3 offer both opportunities and challenges for enhancing the science that EPA produces and applying it to the increasingly complex decisions that are necessitated by wicked problems.

The agency has shown an ability to evaluate new tools and integrate them into its activities in some instances, as described in Chapter 3 and Appendixes C and D, although the process has not been systematic or agency-wide. Also, the agency has made strides in recent years to reorganize and reorient its science activities in ORD with some success. The work of ORD scientists is often the most visible, and at times controversial, scientific interpretation and application in the agency. However, more than three-fourths of the scientific staff in EPA do not work within ORD (EPA SAB 2012b); these scientists are frequently placed in positions where they must apply and interpret science for equally controversial decisions and must be able to access and understand the latest scientific techniques and approaches. There has been progress toward agency-wide science integration with the establishment of the Office of the Science Advisor, and further progress might be made with the shift of the science advisor position from within ORD to the Office of the Administrator in early 2012; however, the Office of the Science Advisor may need further authority from the administrator or additional staff resources to continue to improve the integration and coordination of science across the programs and regions throughout the agency.

As discussed in several places in this report, EPA has made important progress in human health and environmental science and engineering over the last few decades, and the environment is better today because of that progress. However, as the committee reviewed emerging challenges and scientific tools and evaluated the capacity of the agency to respond, the need for substantially enhanced science leadership throughout the agency became clear. When the committee speaks of *enhanced* science leadership, it is not just referring to the strengthened capacity of someone in a high-level position within EPA to whom the administrator has provided independence, authority, and resources, but also the internal support at all levels in the agency (including scientists, analysts, directors, and deputy and assistant administrators) to ensure that the highest-quality science is developed, evaluated, and applied systematically throughout the agency's programs.

At least four independent reports in the last 20 years (EPA 1992, NRC 2000, GAO 2011, EPA SAB 2012b) have, on the basis of their own analyses recommended enhanced science leadership. Some of the specific recommendations included the need for the position of deputy administrator for science with sufficient resources and authority to coordinate scientific efforts in the agency (as noted above) and to build collaboration with external agencies and expertise; the establishment of an overarching issue-based planning process and a scientific agenda for major environmental issues that integrates and coordinates scientific efforts throughout the agency and that is regularly reviewed and updated; a coordinated approach to managing and strengthening EPA's scientific workforce that will serve as a resource for the entire agency; and a strategy that promotes science integration by making it a more consistent priority, by strengthening management oversight, and by strengthening participation and support of EPA scientists. Most recently, the EPA Science Advisory Board (SAB) noted that

> Narrow interpretations of legislative mandates and the organizational structure of the EPA's regulatory programs have posed barriers, in many cases, to innovation and cross-program problem solving. EPA managers and staff in many interviews, especially in program offices, defined the success of their programs in terms of meeting statutory requirements and court-ordered deadlines. Although meeting legal mandates is essential, the EPA needs a broader perspective that extends beyond specific program objectives to achieve multiple environmental protection goals, including

sustainability. A narrow focus on "program silos" and defensibility can be a barrier to formulating and responding to problems as they occur in the real world. Such a limited approach can hinder integration of new scientific information into decisions and new applications of science to develop innovative, effective solutions to environmental problems (EPA SAB 2012b, p. 5).

In the committee's analysis of the strengths and limitations of an enhanced agency-wide leadership position, it has concluded that successful implementation of the systems-based application of emerging tools and technologies to meet persistent and future challenges cannot be achieved under the current structure. Success will require leadership throughout the agency, in the programs and regions as well as in ORD. There will need to be clear lines of authority and responsibility, and regional administrators, program assistant administrators, and staff members at all levels will need to be held accountable for ensuring scientific quality and the integration of individual science activities into broader efforts across the agency.

Finding: The need for improvement in the oversight, coordination, and management of agency-wide science has been documented in studies by the National Research Council, The Government Accountability Office, and the agency's own SAB as a serious shortcoming and it remains an obstacle at EPA. The committee's own analysis of challenges and opportunities for the agency indicates that the need for integration of systems thinking and the need for enhanced leadership at all levels is even stronger than it has been in the past.

Recommendation: The committee recommends that the EPA administrator continue to identify ways to substantially enhance the responsibilities of a person in an *agency-wide* science leadership position. That person should hold a senior position, which could be that of a deputy administrator for science, a chief scientist, or possibly a substantially strengthened version of the current science advisor position. He or she should have sufficient authority and staff resources to improve the integration and coordination of science across the agency. If this enhanced leadership position is to be successful, strengthened leadership is needed throughout the agency and the improved use of science at EPA will need to be carried out by staff at all levels.

Whatever administrative arrangement is adopted, the following are suggestions of the types of responsibilities that the committee thinks should be associated with this position:

- Chairing and assuring that the work of the Science and Technology Policy Council is comprehensive and effective.

- Promotion of systems thinking and systems-oriented tools to address complex challenges ahead and the integration of this approach into every aspect of agency science and engineering (as described in Chapter 4).
- Ensuring that the scientific and technical staff throughout the agency (including program, regional, and research offices) have the expertise necessary to perform their duties whether in support of the agency's research or in support of its role as a regulatory and policy decision-maker.
- Ensuring that the agency has in place a system for quality assurance and quality control of its scientific and technical work (including a system for consistent high-quality peer review).
- Ensuring that the best available scientific and technical information is being used to carry out the agency's mission.
- Working to coordinate research and analytic efforts within and outside the agency to ensure that the best information is used in the most efficient manner.
- Encouraging and supporting interoffice and interagency science collaboration in order to solve problems and develop good solutions.

If the occupant of the position is to be successful, he or she will require sufficient staff and resources to act on behalf of the administrator to implement a coordinated budget and strategic planning process of the regional, program, and research offices to ensure that appropriate scientific and technical expertise and capabilities are available and used. The person in this position would also oversee the policies and procedures related to the operation of the agency's federal advisory committees. **The committee specifically recommends that the person in this position and his or her staff create, implement, and periodically update an integrated,** *agency-wide* **multiyear plan for science, its use, and associated research needs.** Such a plan would bring together ORD, program, and regional science initiatives while being cognizant of the flexibility that is imparted through bottom-up initiatives undertaken in ORD, the program offices, and the regions.

The strengthening of science leadership is not without its challenges. For example, whether or not the position is held by a political appointee could affect the ability of the person in the position to be effective throughout the agency, especially with the other political appointees who head the programs that rely on science (and supervise many of the agency's scientists). There is also the possibility that new procedures established from the central administration could serve to discourage innovation in science if not carefully applied. To a certain extent, the recent EPA decision to re-establish the position of science advisor as a non-political position distinct from ORD (as had been the case in earlier EPA administrations) will provide a test of how to overcome some of these challenges. However, the revised role of the current science advisor does not fully implement the committee's recommendation unless that person is empowered with the tools and support described above. Even with the full support of the administrator and senior staff, the effort will fail if the need to improve the use of science in EPA is not accepted by staff at all levels.

REALIGNMENT OF THE OFFICE OF RESEARCH AND DEVELOPMENT

ORD often sets the stage for research and scientific assessment efforts throughout EPA. In 2011, the deputy administrator for ORD, Paul Anastas, announced a restructuring of the office in response to growing scientific challenges and recommendations from the agency's scientific advisers. The SAB called for integrated transdisciplinary research at ORD, stating that "it will be essential for EPA as a whole, and not just ORD alone, to adopt a systems approach to research planning. It will also be essential to plan and conduct research in new, integrated, and cross-discipline ways to support this systems approach" (EPA SAB 2010). The ORD restructuring aims to

- Align ORD's research with the agency's strategic goals.
- Reorient ORD's research to be guided by the concept of sustainability.
- Promote systems thinking and innovation.
- Couple excellence in problem assessment with excellence in solving problems.
- Encourage integrated, transdisciplinary research among ORD labs and through external funding.

The realignment consolidates 13 previous research sectors into four cross-cutting sectors of research and two overarching sectors, as shown in Table 5-1.

In October 2011, SAB and ORD's Board of Scientific Counselors (BOSC) published a review of ORD's structure (EPA SAB/BOSC 2011). SAB and BOSC noted the "impressive increase in transdisciplinary collaboration as well as coordination across ORD programs with the restructuring." They also made note of ORD efforts to think about innovation operationally as a fundamental aspect of ORD research. SAB and BOSC gave ORD particular credit for having involved regional and program offices in designing the realignment and for giving serious consideration to ways of encouraging creativity among ORD scientists and engineers (EPA SAB/BOSC 2011).

Several key conclusions emerged from the SAB and BOSC review, including suggestions that

- EPA ensure that financial and staff resources are adequate to implement the restructuring and are secured to sustain the communication, stakeholder involvement, and integrated transdisciplinary collaboration that will be essential for its success.

TABLE 5-1 Former and Realigned Structures of EPA's Office of Research and Development

Former ORD Research Structure	Integrated ORD Research Structure
Global Change Research	Air, Climate & Energy
Sustainability Research	
Clean Air Research	
Human Health and Ecosystems Research	
Drinking Water Research	Safe and Sustainable Water Resources
Water Quality Research	
Human Health and Ecosystems Research	Sustainable and Healthy Communities
Pesticides & Toxics Research	
Sustainability Research	
Fellowships	
Land Research (Excluding Nanotechnology)	
Endocrine Disrupting Chemicals Research	Chemical Safety for Sustainability
Computational Toxicology Research	
Human Health & Ecosystems Research	
Human Health Risk Assessment (NexGen)	
Pesticides & Toxics Research	
Land Research (Nanotechnology)	
Clean Air Research (Nanotechnology)	
Sustainability Research	
Human Health Risk Assessment	Human Health Risk Assessment
Homeland Security	Homeland Security

Source: Teichman and Anastas 2011.

- EPA continue to refine its implementation plans to ensure that the restructuring takes root. The agency needs to define clearly how ORD and program office research programs relate to one another and how they fit within the larger context of EPA and stakeholder science. A key aspect is ensuring that senior and junior scientists in ORD and the program offices are invested in the restructuring process.
- EPA develop clear metrics for the evaluation of progress of research divisions and their ability to respond to environmental challenges in a new and more solutions-oriented way. The long-term sustainability of the revised structure (in time and through administrations) will depend on the degree to which the agency can demonstrate that the reorganization leads to better science and better outcomes.

- ORD maintain close communication and working relationships with program offices to ensure that research in the agency continues to support programmatic needs. Regional and program offices should be engaged in evaluating ORD's progress and performance.

COORDINATION OF SCIENCE EFFORTS IN THE US ENVIRONMENTAL PROTECTION AGENCY

The importance of delivering science to EPA decision-makers and supporting the scientific capacities and endeavors of program and regional offices is well-recognized in the agency. The agency should use scientific information in all its decisions. Science needs for decisions are identified within program and regional offices through various processes and can take two main forms—summaries and syntheses of *existing* science and the creation of *new* science to fill key gaps.

Existing science to inform and support decisions is usually acquired by EPA scientific staff (through a combination of professional networks and electronic tools). ORD's Office of Science Policy (OSP) is charged with integrating and communicating scientific information that comes from or that supports ORD's laboratories and centers (EPA 2012a). OSP's Regional Science Program links ORD science to regional offices. The Regional Science Program's Regional Science Liaison and Superfund and Technology Liaison locate scientists in regional offices to facilitate regional staff and management access to ORD science. The regional liaisons have regular communication with OSP to ensure communication between ORD and the regional offices (M. Dannel, EPA, personal communication, December 30, 2011). The EPA SAB Committee on Science Integration for Decision Making found that regional offices consider the liaisons to be important in science acquisition (EPA SAB 2012b).

OSP plays a key role in connecting program and regional offices to ORD research and in expanding the capacity of regional offices to conduct needed research. For a few programs, most notably several programs in the Office of Pesticide Programs, needed research can be required of regulated entities. However, that option is not available to most programs, and those programs and regions rely to various degrees on inhouse research. At the regional level, there are several mechanisms through which new science is supported. For example, the Regional Applied Research Effort Program, which allocates about $200,000 per year to each EPA region for collaborative research, funds near-term research (1–2 years) on high-priority, regional applied-science needs. It is also intended to foster collaboration between EPA regions and ORD laboratories and centers, to build a network between regions and ORD for future scientific interaction, and to provide opportunities for ORD scientists to apply their expertise to regional issues and explore new research challenges. The Regional Methods Program, for which about $600,000 per year is allocated, works to develop new monitoring and enforcement methods (EPA 2012b). It is analogous to the Regional Applied

Research Effort Program in that it provides the regions with near-term research support on high-priority, region-specific science needs and improves collaboration between regions and ORD laboratories and centers (EPA 2008). An example is EPA Region 8, where scientists used support from the Regional Methods Program to collaborate with EPA in developing a vitellogenin gene-induction method to produce a marker of exposure to endocrine disrupting chemicals (Keteles 2011). The Regional Research Partnership Program provides short-term training opportunities (up to 6 months) for regional technical staff to work directly with ORD scientists in ORD laboratories and centers. Regional Science Topic Workshops are held on high-priority topics, including green chemistry, water reuse, and children's environmental exposures. The workshops are intended to identify research needs, initiate research partnerships, and improve information-sharing and coordination of existing research efforts. Through the Regional Research Partnership Program, OSP provides travel and relocation expenses for 10 regional scientists a year to be detailed to specific ORD laboratories for 4 to 12 weeks to work on high-priority research projects in direct collaboration with ORD scientists. The committee concludes that the Regional Science Program could improve the effectiveness of its delivery of ORD and program-office research to regional programs through additional liaisons with specific responsibility in this regard.

ORD is beginning to use social networking and information technology tools, as noted in Chapters 3 and 4 and Appendix D, to promote the development of science communities that cross internal organizational boundaries and extend outside the agency. For example, EPA SAB (2012b) found various electronic sources that are considered useful by the program and regional offices, including the Office of Solid Waste and Emergency Responses CLU-IN Web site (which provides a platform for training, seminars, and podcasts); a variety of forums sponsored by the Office of Solid Waste and Emergency Responses that support the Superfund and Resource Conservation and Recovery Act programs; the Economics Forum, hosted by EPA's National Center for Environmental Economics (NCEE), to keep the agency and other interested parties informed about research; and the Environmental Science Connector, a Web-based tool designed for project management and information-sharing with EPA researchers and external collaborators. ORD is also experimenting with a Web-based collaborative platform called IdeaScale that allows its scientists and engineers to engage in an open, interactive conversation. Staff can share their ideas, then harness the input of their peers through online discussions and ranking tools to refine them. EPA is also developing IdeaScale sites for research programs, engaging both internal and external stakeholders to help in preparing new research frameworks. It is an interesting new approach, but there is little evidence that it has worked effectively to date, having had few users.

Despite the variety of efforts to support and coordinate science within the agency more effectively, the efforts focus on one-way interaction between ORD and program offices or regions and, as noted in several reviews, are not thoroughly coordinated. EPA SAB (2012b, p.7) noted, "ORD principally focuses on

ORD scientists, although it supports several small but important programs in the regions. . . . Program and regional offices manage their scientific workforces relatively independently, with some organizations providing stronger support than others." Given the need for integrated, transdisciplinary, and solutions-oriented research to solve 21st century environmental problems, the existing structure focused on ORD as *the* "science center" that establishes the scientific agenda of EPA will not be sufficient; ORD only makes up a portion of EPA's scientific efforts, and more than three-fourths of EPA's scientific staff work outside ORD (EPA SAB 2012b). When science integration or collaboration occurs, it involves largely short-term needs and problems. Although ORD has surveyed regional and program offices for science and data needs and it will be necessary to continue to conduct regular and systematic assessments of regional and program offices to inform its planning, the focus on ORD planning alone will not be adequate to address science needs for 21st century challenges. As noted above, the development of strategic, coordinated multiyear agency-wide science integration plans, overseen by enhanced science leadership empowered by the administrator, are critical for the agency to coordinate and deliver science in and outside of the agency more effectively in the future. Such integrated plans would also assist the agency in determining where resources outside the agency may be used.

STRENGTHENING SCIENCE CAPACITY

Science flourishes where scientists flourish, and scientists flourish where they have opportunities to work on interesting, challenging problems, interact synergistically with colleagues, have an impact, and earn recognition for their work. In seeking to strengthen its science capacity, EPA needs to attend to the structure of its research operations; to attract, retain, and develop scientific talent within the agency; to contribute to environmental-education efforts to build the talent pool for the future; to support science outside the agency; and to ensure that science is conducted with the utmost integrity. Those points are addressed below.

Enhancing Expertise in the US Environmental Protection Agency

As discussed in Chapters 2 and 3, EPA will need to continue to be prepared to address a wide array of environmental and health challenges and their complex interactions. In some cases, the agency will need to advance scientific understanding through inhouse research efforts; in others, it will need to assimilate and influence scientific efforts that are undertaken elsewhere. Strategic workforce planning when hiring new staff will help to ensure that EPA has expertise it needs in critical fields. Equally important, EPA should carefully attend to the challenge of continuing science education to ensure that scientists are productive throughout their careers even as the pace of change in scientific tools, techniques, and challenges increases.

Building and enhancing capacity of young scientists to be innovators, collaborators, and systems thinkers with a transdisciplinary perspective will require strong leadership, flexibility, and coordination. Given that a large percentage of EPA scientists in ORD and other program offices are near retirement, it is critical for the agency to recruit a new generation of scientists who are well versed in emerging tools (discussed in Chapter 3 and Appendixes C and D) and in cross-disciplinary collaboration and who have been mentored by current scientists. Mentoring will allow younger scientists to gain an understanding of years of research and regulatory science from older scientists. One specific example is in the field of statistics. Senior statisticians are important in EPA because they have the knowledge and experience to mentor inhouse junior statisticians and scientists, facilitate inhouse data analytic work, steer the agency to secure appropriate expert support from outside, and ensure the quality of agency's statistical work. The best type of person to fill this senior position not only has advanced statistical expertise, but also has substantive knowledge in other fields and substantial teamwork experience.

To develop career paths and increase productivity of its newer scientists, EPA needs to be vigilant in engaging them and fostering their professional development. The committee supports ORD's efforts to clarify requirements for promotion of scientists and engineers to senior levels (Anastas 2011). The promotion criteria require substantial achievement that displays high scientific quality, relevance to EPA's mission, and impacts on decision-making. As is typical of expectations in most academic institutions, scientists and engineers seeking promotion to the GS-14 level are expected to be nationally recognized for their contributions and those seeking promotion to GS-15 to have international recognition. ORD's promotion criteria now highlight expectations for transdisciplinary research, teamwork, and leadership (Anastas 2011).

EPA also needs larger and more senior cadres of scientists in fields in which it wants to play a strong leading role among federal agencies (NRC 2010a). In a recent example, EPA's National Center for Computational Toxicology (NCCT) was established to address the lack of toxicity data on the many chemicals that are on the market and to do so in an efficient and cost-effective manner (see Chapter 3 and Appendix C for more information about EPA's computational toxicology program). Buoyed by the guidance and affirmation it received from *Toxicity Testing in the 21st Century* (NRC 2007), ORD and NCCT leadership set an ambitious path to address their charge. In its first 5 years, the center has been able to break boundaries and build transdisciplinary collaborations with other federal partners and the private sector both in the United States and internationally. The science generated through the center's collaborations has created momentum around computational toxicology research and influenced research investments by other agencies and organizations, including the chemical industry.

Optimizing resources, creating and benefiting from scientific exchange zones, and leading innovation through transdisciplinary collaborations to address the many challenges described in Chapter 2 will require forward-thinking and resourceful scientific leadership at various levels in the agency. This includes EPA using all of its authority effectively, including pursuing permanent Title 42 authority, to recruit, hire, and retain the high-level science and engineering leaders that it needs to maintain a strong inhouse research program (NRC 2010a). It will also mean maintaining a critical mass of world-class experts who have the ability to identify and access the necessary science inside or outside EPA and to work collaboratively with researchers in other agencies.

Finding: Expertise in traditional scientific disciplines—including but not limited to statistics, chemistry, economics, environmental engineering, ecology, toxicology, epidemiology, exposure science, and risk assessment—are essential for addressing the challenges of today and the future. The case of statistics is one example where the agency is facing significant retirements and needs to have, if anything, enhanced expertise.[1] EPA is currently attuned to these needs, but staffing high-quality scientists in these areas of expertise who can embrace problems by drawing from information across disciplines will require continued attention if EPA is to maintain its leadership role in environmental science and technology.

Recommendation: EPA should continue to cultivate a scientific workforce across the agency (including ORD, program offices, and regions) that can take on transdisciplinary challenges.

The committee recognizes that EPA already provides many unique opportunities to engage in high quality, collaborative, and interdisciplinary research. However, EPA can continue to build its capacity by cultivating a scientific workforce across the agency (including ORD, program offices, and regions) that can take on transdisciplinary challenges. Some options that EPA might explore to fulfill the recommendation above include:

- Build a stronger mentoring and leadership development program that supports young researchers and fosters the culture of systems-thinking research.
- Recruit young scientists who have expertise and interest in scientific concepts and tools relevant to systems thinking.
- Promote rotations through its laboratories and through the laboratories of other federal agencies and scientific organizations as valuable training experiences for new scientists in the areas of environmental health, science, and engineering.

---

[1] ORD currently has 12 epidemiologists, 31 statisticians (mathematical or research), and 8 biologic and health statisticians (E. Struble, EPA, personal communication, July 13, 2012). These job titles typically require a certain amount of statistics course work and do not fully reflect statistical expertise across the entire agency. There are staff members with other job titles who also fulfill the data analysis role.

- Expand opportunities for internal networking, including opportunities for scientists and engineers to work between programs and offices.
- Encourage scientists and engineers to work in interdisciplinary teams and in new ways to provide information in a timely fashion.
- Implement programs to help scientists and engineers to acquire new skills and expertise throughout their careers, including educational opportunities, sabbaticals and other kinds of leave, and laboratory rotations.
- Provide opportunities for agency scientists to interact with colleagues in other agencies, in universities, in non-profit organizations, and in the private sector; such opportunities could include workshops, roundtables, participation in traditional research conferences, and longer-term exchanges with or as visiting scientists.
- Promote the visibility and recognition of scientific excellence across its divisions, programs, and locations by enhancing and highlighting its featured research and awards programs.
- Assess its current policies for retaining and hiring civil service employees. The agency should be nimble and should be able to easily hire or reassign employees to make sure it has specific expertise to understand emerging challenges and make use of new tools, technologies, and approaches in the appropriate offices, regions, and laboratories at the appropriate time.

There are several fields in which EPA lacks expertise and in which investments in additional expertise could provide substantial benefits to the agency and its mission. One key recognized need is for social, behavioral, and decision scientists. EPA's economic, social, behavioral, and decision science staff consists almost entirely of economists. The agency is without strong expertise in social, behavioral, and decision sciences, though it does support some research in these areas through the Science to Achieve Results (STAR) program and procures economics research from contractors. Social research in EPA was historically funded by ORD (NRC 2000). In 2008, the economics and decision science extramural research program was transferred to NCEE (EPA SAB 2011). As part of the reorganization, decision sciences were eliminated altogether (EPA SAB 2011). Economics has remained a low priority (EPA SAB 2011) and the economics staff (about 100 economists) is a very small fraction of the agency's professional staff (EPA, unpublished material, 2012[2]).

The small representation of economics expertise and the virtual nonexistence of behavioral and decision scientists (nine social scientists, four psychologists, and one sociologist; [3][P. Vaughn, EPA, personal communication, July 13,

---

[2]The unpublished data was received from M. Bender, EPA, on July 13, 2012, as part of a data request made on behalf of the Committee on Science for EPA's Future. This material is available by contacting the Public Access Records Office of the National Academies.

[3]In addition to psychologists and sociologists, NCEE acknowledged that it was lacking expertise in behavioral economics and is pursuing the hiring of new staff in that field.

2012] in the professional staff are matters of concern, given EPA's ever-present and increasing need to defend programs and initiatives on economic grounds, its concerns for environmental justice and community engagement in environmental decision-making, and its goal of becoming more transdisciplinary. In addition, and as noted above, economic, social, behavioral, and decision sciences can make important contributions to improving environmental policy decisions within the emerging integrated systems-based approach to environmental management and contributions to innovation in strategies for achieving environmental goals efficiently, equitably, and cooperatively. The importance of behavioral science to the conduct of environmental economics research, including environmental valuation, has been well established and has led to considerable research on the integration of behavioral sciences with environmental economics (Sent 2004; Shogren and Taylor 2008; Shogren et al. 2010).

In particular, behavioral and decision science deals with such issues as "framing effects" and the role of cognitive heuristics (see, for example, Kahneman et al. 1982). It provides the intellectual basis of modern methods of risk communication (see, for example, Morgan et al. 2002), and its insights and ideas should constitute a key complement of economics and decision analysis in such contexts as the design and assessment of the likely effectiveness of alternative regulatory strategies (Fischbeck and Farrow 2001). Those insights often arise from extensive experimental studies in both laboratory and field settings. Without staff in EPA who are experts in those subjects, not only can the agency not conduct such studies, but often it does not know how to ask the right questions or how to seek the right expertise and advice. That is unfortunate—social, behavioral, and decision sciences are essential in the design of survey research, the development of methods in expert elicitation to characterize the state of uncertain science (EPA 2009a), the development and evaluation of risk-communication methods, the development and use of mediation and other group-decision processes, the promotion of a variety of environmentally benign behaviors, and the anticipation of behavioral responses to alterative regulatory and other protective strategies. EPA needs to have staff with sufficient expertise or cross-disciplinary training to allow it to become an educated consumer of social, behavioral, and decisions sciences and to engage more effectively with external entities to conduct innovative science.

Finding: EPA's economic, social, behavioral, and decision science staff consists almost entirely of economists. The agency is without strong expertise in social, behavioral, and decision sciences, though it does support some research in these areas through outside grants, collaborations, and procurement.

Recommendation: The committee recommends that EPA add staff who have training in behavioral and decision sciences and find ways to enhance the existing staff capabilities in these fields.

Options that EPA might explore to fulfill that recommendation include:

- Recruit several new staff who have earned advanced degrees in empirically-based behavioral and decision science. The new staff would need to have strong communication skills and would need to work with economists, natural scientists, and engineers in the agency to help to make regulatory and other agency policies that promote environmentally protective behaviors and that are more realistic. Their knowledge would assist the agency by helping it to make more informed choices in seeking outside contractors and advisers and to create stronger collaboration with academics in related fields. The committee suggests that the new staff be located within NCEE. The reason for that suggestion is that NCEE currently staffs the largest number of social scientists within the agency. The large interest in behavioral and decision sciences that exists now in economics broadly, as exhibited by the fields of behavioral and neuroeconomics, will contribute to making NCEE a productive location. More importantly, behavioral economics is an essential source of new insight in environmental economics research pertaining to the benefits of environmental protection and the design of incentives for environmental management. Co-locating behavioral scientists within NCEE will increase the capacity of economics staff to participate in the advances in environmental economics emerging from the integration of behavioral economics.
- Provide mechanisms for cross-disciplinary training of staff in core disciplines relevant to behavioral and decision science. The committee acknowledges that the number of staff in EPA who have advanced training in these fields is likely to remain modest even with a concerted recruitment effort, and it is important for staff scientists who work in adjacent disciplines to have enough familiarity to know what questions to ask (and whom to ask).
- Develop improved mechanisms for integrating economic, social, behavioral, and decisions sciences into the development of science to support environmental-management decisions.

Using Outside Expertise

EPA often needs to seek expertise and research from sources outside the agency when science needs cannot be met from within. Sources include other federal agencies (such as, the Department of Defense, the Department of Energy, the National Oceanic and Atmospheric Administration, and the US Department of Agriculture), government research organizations (such as the National Institute of Environmental Health Sciences), industrial research initiatives, universities, consultants, state and local governments, and nongovernment organizations (EPA SAB 2011). The international community is also a good resource for EPA, as there is a lot of high-quality environmental and human health research undertaken in Europe, Asia, and elsewhere. International collaboration is particularly important considering products and processes are becoming more global and considering many environmental problems, such as the transport of pollutants, are global in nature.

Strategic collaborations with other agencies and scientific institutions will be critical if EPA is to access the breadth of expertise necessary to address 21st century environmental challenges. For example, chemical and pesticide regulations are informed by hazard data derived from animal toxicologic studies. But as epidemiologic and biomonitoring studies generate more information that is relevant to risk assessment, a broader array of expertise will be required to interpret the new types of data and weigh their evidence relative to the more prevalent toxicologic data. EPA needs to have sufficient internal expertise and critical mass in epidemiology, biostatistics, and population-based research. However, rather than house large teams of epidemiologists and biostatisticians among its experts, EPA could build collaborative networks with the National Institute of Environmental Health Sciences and other agencies to undertake assessments. In fields in which it is unrealistic to have sufficient inhouse capacity, existing scientific staff at EPA will need to have adequate cross-disciplinary awareness to ask the right questions and identify appropriate collaborators. For example, if statistical expertise is needed from an outside source, the contractor or subcontractor that is hired should have adequate expertise in statistics (such as a PhD) to successfully meet EPA's needs.

Building New Expertise Through Education

The future of EPA's scientific enterprise depends on having a diverse body of capable and committed scientists and engineers to work in EPA and in research positions in other government agencies, academe, the nonprofit sector, and the private sector. Future scientists and engineers should understand the complex nature of environmental challenges and the transdisciplinary needs and opportunities for solutions. Furthermore, to achieve its mission of protecting human health and the environment, the agency will need to play a role in helping to educate and engage the public. Public understanding and engagement are especially critical in EPA's efforts to achieve its aims by using nonregulatory approaches and in building ongoing support for the environmental science and engineering and protection efforts of the agency. Among other needs, the agency will need to educate stakeholders and the public about new scientific concepts and approaches that it develops or adopts, and to provide training for potential users of new tools and technologies. EPA has numerous valuable programs that are designed to increase the pipeline of future environmental engineers and scientists and to expand and improve environmental education more broadly. Early environmental education is important in creating champions for environmental protection and innovation in new science and technology who can work in the agency in the future.

The National Environmental Education Act of 1990 simultaneously established the Office of Environmental Education (OEE) in EPA and the National Environmental Education Foundation, a nonprofit corporation meant to leverage private support. The act authorized environmental-education grants, internship and fellowship programs, and the Environmental Education and Training Partnership, which has worked to develop standards for environmental education. In 2009, OEE issued an *Environmental Education Highlights* report (EPA 2009b) that briefly describes some of the dozens of outreach and education programs that EPA leads. They include collaborations with schools, the Boys and Girls Clubs, the Girl Scouts, and the Parent–Teacher Organization to provide education and service opportunities focused on energy conservation, water conservation, recycling, and waste reduction. The Tools for Schools program in the Office of Air and Radiation has reached more than 60,000 schools with educational materials, training, and guidance on indoor air quality. The agency collaborates with the American Meteorological Society to provide training and outreach tools for broadcast meteorologists on air quality and watershed protection. Between scientific survey trips, EPA's ocean survey vessel *Bold* hosts open houses at ports of call around the country.

OEE administers the National Network for Environmental Management Studies fellowships, which were established in 1986 and have supported more than 1,400 fellows. Network fellows receive support for undergraduate or graduate studies and work on EPA-supported and EPA-directed research projects. Students apply in response to requests for applications developed by EPA staff in Washington, DC, and in regional offices and laboratories around the country.

ORD also offers critical student support and encouragement through its People, Prosperity, and the Planet (P3) student grants and design competition, its Greater Research Opportunities undergraduate fellowship program, and the STAR graduate fellowship program (EPA 2012c). The Greater Research Opportunities program offers fellowships to juniors and seniors who are studying in environment-related fields in colleges and universities that do not receive large amounts of federal research funding. The fellowships provide academic support for up 2 years with a summer internship at EPA. The agency plans to award about $2 million worth of Greater Research Opportunities undergraduate fellowships in 2012 (EPA 2012d). The STAR fellowship program supports master's and doctoral students who are working in environment-related fields. Students competing for STAR grants are required to submit original proposals on EPA-specified research topics that run the gamut from social sciences to engineering. More than 1,500 STAR fellowships have been awarded since the program began in 1995 (EPA 2012c). The P3 program offers grants to teams of college students who research and design innovative solutions to sustainability challenges (EPA 2012e). The teams can apply for $15,000 for Phase 1 development grants and up to $90,000 in Phase 2. Phase 2 grants are awarded at the National Sustainable Design Exposition in Washington, DC, each April (EPA 2011a).

Delivering Science Outside the Environmental Protection Agency

As state, local, and tribal environmental agency budgets decline, the agencies will rely increasingly on EPA for scientific support. EPA conducts programs that are intended to provide and communicate science and tools for decision-makers and practitioners outside EPA. Several of EPA's large-scale regional research programs (for example, the Chesapeake Bay, Great Lakes, and Puget Sound programs) are designed specifically to develop and deliver science and decision support tools to help environmental authorities outside EPA. ORD conducts research programs to develop widely applicable decision-support tools. ORD's Collaborative Science and Technology Network for Sustainability provides grants to explore "new approaches to environmental protection that are systems-oriented, forward-looking, preventive, and collaborative" (EPA 2011b). The Tribal Science Program supports community-based research in an effort to improve understanding of the relationship between tribal-specific factors and health risks posed by toxic substances in the environment (EPA 2011c). Web-based platforms are essential for delivering science and tools to state, local, tribal, and other non-EPA practitioners, and EPA has made an effort to take advantage of such platforms, such as the Health and Environmental Research Online (HERO) system (see below).

INTEGRITY, ETHICS, AND TRANSPARENCY IN THE US ENVIRONMENTAL PROTECTION AGENCY'S PRODUCTION AND USE OF SCIENTIFIC INFORMATION

Since its founding, EPA has been challenged by the need to use the best available scientific information in developing policy and regulations. Critics of EPA's regulations (as either too lax or too stringent) have sometimes charged that valid scientific information was ignored or suppressed, or that the scientific basis of a regulation was not adequate. EPA's best defense against those criticisms is to ensure that it transparently distinguishes between questions of science and questions of policy in its regulatory decisions; to demand openness and access to the scientific data and information on which it is relying, whether generated in or outside of the agency; and to use competent, balanced, objective, and transparent procedures for selecting and weighing scientific studies, for ensuring study quality, and for peer review.

Distinguishing Science Questions from Policy Questions

In a memorandum on scientific integrity issued on March 9, 2009, President Obama declared that "political officials should not suppress or alter scientific or technologic findings and conclusions" (The White House 2009). After the president's directive, EPA administrator Lisa Jackson stated in a memo issued on May 9, 2009, that

> while the laws that EPA implements leave room for policy judgments, the scientific findings on which these judgments are based should be arrived at independently using well-established scientific methods, including peer

review, to ensure rigor, accuracy, and impartiality. This means that policymakers must respect the expertise and independence of the Agency's career scientists and independent advisors while insisting that the Agency's scientific processes meet the highest standards of rigor, quality, and integrity (Jackson 2009).

The Bipartisan Policy Center (2009) has recommended that the best means for regulators to reduce opportunities for inappropriate political intervention in scientific judgments and to avoid the perception that politicization of science had occurred is to distinguish clearly between science and policy questions in formal regulatory documents. EPA has done that well in recent reviews of the National Ambient Air Quality Standards, which separate the review of scientific information on health and welfare effects presented in its integrated science assessments from the policy-assessment documents that draw on the scientific information. Continting to promote that approach will support the distinction between science questions and policy questions and conducting periodic audits of rulemaking documents will help to ensure compliance with the distinctions. Given the uncertainties surrounding many complex environmental problems, it is important for the agency to be transparent about types of uncertainties involved in its assessments and to be clear about how both science and policy considerations inform ultimate decisions.

Increased Access to Scientific Information

One of the key elements of ensuring the credibility of science used in decision-making is maintaining the highest level of transparency, and making scientific information used in EPA decisions easily accessible, as much as possible, to all parties who are interested. That access includes

- *Access to the full array of published scientific evidence.* One example of important progress has been the recent development by EPA of a searchable electronic database, the HERO system, to give its own staff full access to the emerging scientific literature and give the public access to a searchable on-line database of citations of all studies reviewed in support of its regulations. This is a valuable tool that should continue to have support.
- *Access to data.* When regulatory stakeholders have legitimate interest in examining the data that underlie reported results, access to published articles or reports is not sufficient. Since the late 1990s, the Data Access Act (or Shelby Amendment) has required that recipients of federal research funding provide their research data to requesting parties if the federal government has used their research findings in developing regulations and the data are later requested under the Freedom of Information Act (OMB Circular A-110). That requirement allows requesting parties the opportunity to inspect and reanalyze data that were

used to support regulations. The Office of Management and Budget (OMB) circular that contains the requirement exempts preliminary analyses, drafts of papers, plans for future research, peer reviews, and communications with colleagues. It also exempts trade secrets, commercial information, and information that must be withheld to protect the privacy of research subjects. The Data Access Act is consistent with an interest in providing greater access to scientific information that underlies regulatory efforts but is limited in applying only to federally funded research (Wagner 2003; Wagner and Michaels 2004). It would be useful to extend requirements for data access to privately supported research that is submitted for regulatory purposes. As with publicly supported research, exemptions could be provided as necessary to protect the privacy of research subjects and legitimate proprietary interests.

- *Access to EPA internal research.* Concerns about access apply not only to externally sourced scientific information but to research data and findings that are developed through EPA's internal research programs (Grifo 2009). Publication of EPA science not only helps to bolster the agency's influence, it also provides legitimacy in the scientific community. EPA needs to encourage its own scientists to communicate and publish their results and to do so in a timely manner. Institutional barriers to the publication of results, particularly bureaucratic delays related to internal approvals and concerns about policy implications, should be addressed.

Ensuring the Quality of Scientific Information

In rule-making processes that rely on extensive reviews of scientific information, EPA generally imposes a strong preference for reliance on published, peer-reviewed studies. The agency's peer review policy states that "peer review of all scientific and technical information that is intended to inform or support Agency decisions is encouraged and expected" (EPA 2006). The OMB *Final Information Quality Bulletin for Peer Review* (OMB 2004) and EPA's internal *Peer Review Handbook* (EPA 2000) guide the peer-review process for internally generated scientific studies and tools. However, when EPA needs to go beyond peer-reviewed literature to fill information gaps, it may need to be more active in initiating external peer review to ensure that the identified externally generated information is reliable and to provide quality assurance for stakeholders.

EPA has used advisory groups both to review scientific research and to provide advice and expertise from outside the agency. For example, EPA's National Advisory Council for Environmental Policy and Technology (NACEPT) was established in 1988 to use environmental-policy expertise outside the agency. The advisory council is an independent group of experts that has provided advice to EPA on a broad variety of topics, including workforce capacity, strategic planning, promotion of environmental stewardship, and strategies for improving access to environmental information (EPA 2012d). Various other advisory committees, established under the Federal Advisory Committee Act, provide scientific advice on such issues as environmental justice and children's environmental health. A 2009 review of EPA's Office of Cooperative Environmental Management found that although committees like NACEPT were useful tools for the agency, there was a lack of coordination between other committees and agency advisory boards, such as SAB and BOSC (EPA 2009c). External advisory groups—including SAB, BOSC, and NACEPT—play an important role in helping EPA to ensure the credibility and quality of its scientific studies and science-based decisions. They will remain a valuable resource for the agency assuming the members of these bodies continue to be chosen based on the virtue of their expertise and experience and are appropriately tasked with providing advice that falls within the purview of scientific experts.

Even when the underlying science meets the highest standards of quality and integrity, judgment is used to select and weigh studies that will be used for decision-making. EPA has developed various guidelines to weigh studies and evaluate science, such as guidelines developed in response to sections 108 and 109 of the Clean Air Act. However, EPA has sometimes been criticized for its failure to describe clearly its criteria and methods to identify, evaluate, and weigh scientific studies. For example, National Research Council (NRC) reports over the last decade have evaluated health assessments developed for EPA's Integrated Risk Information System (IRIS) and indicated a need to improve formal, evidence-based approaches to increase transparency and clarity for selecting datasets for analysis, and to focus more on uncertainty and variability (NRC 2005, 2006, 2010b).

Many of the above observations were reflected in the *Review of the Environmental Protection Agency's Draft IRIS Assessment of Formaldehyde* (NRC 2011). In its review, the authoring committee of that report noted a lack of clarity and transparency in the methods used to assess the health effects of formaldehyde. Specifically, that committee found the assessment did not contain "sufficient documentation on methods and criteria for identifying evidence from epidemiologic and experimental studies, for critically evaluating individual studies, for assessing the weight of evidence, and for selecting studies for derivation of the [reference concentrations] and unit risk estimates" (NRC 2011). The report made several recommendations that were specific to improving the formaldehyde IRIS assessment, but also provided some suggestions for improving the IRIS process.

Deficiencies in EPA's IRIS assessments have resulted in some critics casting doubt on the science used to support agency decisions. EPA is aware of those stakeholder criticisms and of the problems identified by the NRC (2005, 2006, 2010b, 2011), and it has announced improvements in the IRIS assessments that will be reviewed by the recently assembled NRC Committee to Review the IRIS Process. This example illustrates the need for formal evidence-based approaches that are clearly documented and well-reviewed; they can be protective of EPA's science-informed policies.

STRENGTHENING SCIENCE IN A TIME OF TIGHT BUDGETS

This report has stressed the importance of sustaining and strengthening EPA's present programs of scientific research, applications, and data collection while identifying and pursuing a wide array of new scientific opportunities and challenges. Both are needed to address the complexity of modern problems and both are essential to the agency if it is to continue to provide scientific leadership and high-quality science-based regulation in the years to come.

Specific recommendations related to agency budgets are outside the scope of this study, but the committee feels compelled to note, as did the report *Science Advisory Board Comments on the President's Requested FY2013 Research Budget* (EPA SAB 2012a), that since 2004, the budget for ORD has declined 28.5% in real-dollar terms (gross domestic product–indexed dollars). The reductions have been even greater in a number of specific fields, such as ecosystem research and pollution prevention.

Finding: If EPA is to provide scientific leadership and high-quality science-based regulation in the coming decades, it will need adequate resources to do so. Some of the committee's recommendations, if followed, will allow EPA to address its scientific needs with greater efficiency. But the agency cannot continue to provide leadership, pursue many new needs and opportunities, and lay the foundation for ensuring future health and environmental safety unless the long-term budgetary trend is reversed.

Recommendation: The committee recommends EPA create a process to set priorities for improving the quality of its scientific endeavors over the coming decades. This process should recognize the inevitably limited resources while clearly articulating the level of resources required for the agency to continue to ensure the future health and safety of humans and ecosystems.

SUMMARY

It is clear that if EPA is to meet current, persistent, and future challenges and is to succeed in applying systems thinking throughout its scientific enterprise, it will have to continue to enhance its scientific capacity and improve coordination of science throughout the agency. In this chapter, the committee has described how EPA can enhance its agency-wide science leadership, take steps to continue the realignment of ORD to advance transdisciplinary research and support the agency's strategic goals, strengthen internal scientific capacity and ties to the larger environmental science and engineering research community, and ensure the integrity of the scientific information the agency generates or uses.

REFERENCES

Anastas, P. 2011. Technical Qualifications Board (TQB) Policy-Supplemental Guidance. Memorandum to Laboratory Directors and Center Directors, from Paul Anastas, Assistant Administrator for Research and Development, US Environmental Protection Agency, Washington, DC. June 9, 2011. Attachment: Additional Guidance for Promotion of ORD Scientists and Engineers in Research, Development, and Expert Positions, May 2010.

EPA (US Environmental Protection Agency). 1992. Safeguarding the Future: Credible Science, Credible Decisions: The Report of the Expert Panel of the Role of Science at EPA. EPA/600/9-91-050. US Environmental Protection Agency, Washington, DC. March 1992.

EPA (US Environmental Protection Agency). 2000. Peer Review Handbook. EPA 100-B-00-001. Office of Science Policy, Office of Research and Development, Washington, DC [online]. Available: http://www.epa.gov/peerreview/pdfs/prhandbk.pdf [accessed Apr. 18, 2012].

EPA (US Environmental Protection Agency). 2006. Peer Review and Peer Involvement in the US EPA [online]. Available: http://www.epa.gov/peerreview/pdfs/peer_review_policy_and_memo.pdf [accessed Apr. 12, 2012].

EPA (US Environmental Protection Agency). 2008. Smart Energy Resources Guide. EPA/600/R-08/049. US Environmental Protection Agency [online]. Available: http://www.arta1.com/cms/uploads/Smart%20Energy%20Resource%20Guide.pdf [accessed Apr. 19, 2012].

EPA (US Environmental Protection Agency). 2009a. Expert Elicitation Task Force White Paper, Draft. Science Policy Council, US Environmental Protection Agency, Washington, DC. January 6, 2009 [online]. Available: http://www.epa.gov/osa/pdfs/elicitation/Expert_Elicitation_White_Paper-January_06_2009.pdf [accessed Apr. 12, 2012].

EPA (US Environmental Protection Agency). 2009b. Environmental Education Highlights. US Environmental Protection Agency [online]. Available: http://www.epa.gov/enviroed/pdf/2009_EEHighlights.pdf [accessed Apr. 12, 2012].

EPA (US Environmental Protection Agency). 2009c. 20 Years of Shaping Environmental Policy at EPA. EPA 130-R-09- 003. Office of Cooperative Environment Management, US Environmental Protection Agency [online]. Available: http://nepis.epa.gov/Adobe/PDF/500025SF.PDF [accessed Apr. 12, 2012].

EPA (US Environmental Protection Agency). 2011a. Funding Opportunities. Extramural Research, US Environmental Protection Agency [online]. Available: http://www.epa.gov/ncer/rfa/2012/2012_p3.html [accessed Apr. 19, 2012].

EPA (US Environmental Protection Agency). 2011b. Collaborative Science and Technology Network for Sustainability (CNS), Extramural Research, US Environmental Protection Agency [online]. Available: http://epa.gov/ncer/cns/ [accessed Apr. 19, 2012].

EPA (US Environmental protection Agency). 2011c. Collaborative Community and Regional Programs. US Environmental Protection Agency [online]. Available: http://epa.gov/ncer/cns/programs.html [accessed Apr. 19, 2012].

EPA (US Environmental Protection Agency). 2012a. Science Policy. US Environmental Protection Agency [online]. Available: http://www.epa.gov/osp/ [accessed Apr. 19, 2012].

EPA (US Environmental Protection Agency). 2012b. Regional Methods (RM) Program. US Environmental Protection Agency [online]. Available: http://www.epa.gov/osp/regions/rm.htm [accessed Apr. 18, 2012].

EPA (US Environmental Protection Agency). 2012c. Fellowships. Extramural Research, US Environmental Protection Agency [online]. Available: http://www.epa.gov/ncer/fellow/ [accessed Apr. 18, 2012].

EPA (US Environmental Protection Agency). 2012d. Fall 2012 EPA Greater Research Opportunities (GRO) Fellowships for Undergraduate Environmental Study. Greater Research Opportunities (GRO) Program, Office of Research and Development, US Environmental Protection Agency[online]. Available: http://www.epa.gov/ncer/rfa/2012/2012_gro_undergrad.html [accessed May 17, 2012].

EPA (US Environmental Protection Agency). 2012e. P3: People, Prosperity and the Planet Student Design Competition for Sustainability. US Environmental Protection Agency [online]. Available: http://www.epa.gov/ncer/p3/ [accessed Apr. 18, 2012].

EPA SAB (US Environmental Protection Agency Science Advisory Board). 2010. Office of Research and Development Strategic Research Directions and Integrated Transdiciplinary Research. EPA-SAB-10-010. Memo to Lisa P. Jackson, Administrator, US Environmental Protection Agency, from Deborah L. Swackhamer, Chair, Science Advisory Board, US Environmental Protection Agency, Washington, DC. July 8, 2010 [online]. Available: http://yosemite.epa.gov/sab/sabproduct.nsf/E989ECFC125966428525775B0047BE1A/$File/EPA-SAB-10-010-unsigned.pdf [accessed Apr. 13, 2012].

EPA SAB (US Environmental Protection Agency Science Advisory Board). 2011. Science Advisory Board Comments on The President's Requested FY 2012 Research Budget. EPA-SAB-11-007. Memo to Lisa P. Jackson, Administrator, from Deborah L. Swackhamer, Chair, Science Advisory Board, and Jerold Schnoor, Chair, SAB Research Budget Work Group, US Environmental protection Agency, Washington, DC. June 2, 2011 [online]. Available: http://yosemite.epa.gov/sab/sabproduct.nsf/c91996cd39a82f648525742400690127/9BE9A90F43A8DD1D852578A30069D7E5/$File/EPA-SAB-11-007-unsigned.pdf [accessed Apr. 18, 2012].

EPA SAB (US Environmental Protection Agency Science Advisory Board). 2012a. Science Advisory Board Comments on the President's Requested FY 2013 Research Budget, May 3, 2012. Science Advisory Board, US Environmental Protection Agency, Washington, DC [online]. Available: http://yosemite.epa.gov/sab/sabproduct.nsf/0/1190D2161DBCAD3B852579F3005FC0CF/$File/EPA-SAB-12-006-unsigned-SS.pdf [accessed Aug. 16, 2012].

EPA SAB (US Environmental Protection Agency Science Advisory Board). 2012b. Science Integration for Decision Making at the US Environmental Protection Agency(EPA). Final Report, July 6, 2012. Science Advisory Board, US Environmental Protection Agency, Washington, DC [online]. Available: http://yosemite.epa.gov/sab/sabproduct.nsf/fedrgstr_activites/8AA27AA419B1D41385257A330064A479/$File/EPA-SAB-12-008-unsigned.pdf [accessed July 22, 2012].

EPA SAB/BOSC (US Environmental Protection Agency Science Advisory, and Board and Board of Scientific Counselors). 2011. Office of Research and Development (ORD) New Strategic Research Directions: A Joint Report of the Science Advisory Board (SAB) and ORD Board of Scientific Councilors (BOSC). EPA-SAB-12-001. Science Advisory Board, and Board of Scientific Councilors US Environmental Protection Agency, Washington, DC. October 21, 2011[online]. Available: http://www.epa.gov/osp/bosc/pdf/StratResDir111021rpt.pdf [accessed Apr. 13, 2012].

Fischbeck, P.S., and R.S. Farrow, eds. 2001. Improving Regulation: Cases in Environment, Health, and Safety. Washington, DC: Resources for the Future.

GAO (US Government Accountability Office). 2011. Environmental Protection Agency: To Better Fulfil It's Mission, EPA Needs a More Coordinated Approach to Managing its Laboratories. July 2011. GAO-11-347. Washington, DC: US Government Accountability Office [online]. Available: http://www.gao.gov/assets/330/321850.pdf [accessed April 13, 2012].

Grifo, F. 2009. Testimony of Francesca T. Grifo, Senior Scientist with the Union of Concerned Scientists, Director of the Scientific Integrity Program Before the US Senate Committee on Environment and Public Works "Scientific Integrity and Transparency Reforms at the EPA", June 9, 2009 [online]. Available: http://www.ucsusa.org/scientific_integrity/solutions/agency-specific_solutions/ucs-testimony-to-senate-June-2009.html [accessed Apr. 18, 2012].

Jackson, L.P. 2009. Scientific Integrity. Memo to all EPA Employees, from Lisa P. Jackson, Administrator, US Environmental Protection Agency, Washington, DC. May 9, 2009 [online]. Available: http://blog.epa.gov/administrator/2009/05/12/memo-to-epa-employees-scientific-integrity/ [accessed Apr. 18, 2012].

Kahneman, D., P. Slovic, and A. Tversky. 1982. Judgment under Uncertainty: Heuristics and Biases. Cambridge: Cambridge University Press.

Keteles, K. 2011. Science at EPA: A Regional Office Perspective. Presentation to the Third Meeting on Science for EPA's Future, August 8, 2011, Washington, DC.

Morgan, M.G., B. Fischhoff, A. Bostrom, and C.J. Atman. 2002. Risk Communication: A Mental Models Approach. Cambridge: Cambridge University Press.

NRC (National Research Council). 2000. Strengthening Science at the US Environmental Protection Agency: Research-Management and Peer-Review Practices. Washington, DC: National Academy Press.

NRC (National Research Council). 2005. Health Implications of Perchlorate Ingestion. Washington, DC: National Academies Press.

NRC (National Research Council). 2006. Health Risks from Dioxin and Related Compounds: Evaluation of the EPA Reassessment. Washington, DC: National Academies Press.

NRC (National Research Council). 2007. Toxicity Testing in the 21st Century: A Vision and a Strategy. Washington, DC: National Academies Press.

NRC (National Research Council). 2010a. The Use of Title 42 Authority at the US Environmental Protection Agency: A Letter Report. Washington, DC: National Academies Press.

NRC (National Research Council). 2010b. Review of the Environmental Protection Agency's Draft IRIS Assessment of Tetrachloroethylene. Washington, DC: National Academies Press

NRC (National Research Council). 2011. Review of the Environmental Protection Agency's Draft IRIS Assessment of Formaldehyde. Washington, DC: National Academies Press.

OMB (Office of Management and Budget). 2004. Final Information Quality Bulletin for Peer Review [online]. Available: http://www.whitehouse.gov/sites/default/files/omb/memoranda/fy2005/m05-03.pdf [accessed Apr. 19, 2012].

Sent, E.M. 2004. Behavioral economics: How psychology made its (limited) way back into economics. Hist. Polit. Econ. 36(4):735-760.

Shogren, J.F., and L.O. Taylor. 2008. On behavioral-environmental economics. Rev, Environ. Econ. Policy 2(1):26-44.

Shogren, J.F., G.M. Parkhurst, and P. Banerjee. 2010. Two cheers and a qualm for behavioral environmental economics. Environ. Resour. Econ. 46(2):235-247.
Teichman, K., and P. Anastas. 2011. Science for EPA's Future: Innovative Thinking, Creative Solutions. Presentation at the Second Meeting on Science for EPA's Future, June 17, 2011, Washington, DC.
The Bipartisan Policy Center. 2009. Improving the Use of Science in Regulatory Policy. Science for Policy Project Final report. August 5, 2009 [online]. Available: http://bipartisanpolicy.org/sites/default/files/BPC%20Science%20Report%20fnl.pdf [accessed Apr. 18, 2012].
The White House. 2009. Scientific Integrity. Memorandum for the Heads of Executive Departments and Agencies, from President Obama, May 9, 2009 [online]. Available: http://www.ucsusa.org/assets/documents/scientific_integrity/President-Obama-Scientific-Integrity-Memo.pdf [accessed Apr. 19, 2012].
Wagner, W.E 2003. The "bad science" fiction: Reclaiming the debate over the role of science in public health and environmental regulation. Law Contemp. Probl. 66(Fall):63-124.
Wagner, W., and D. Michaels. 2004. Equal treatment for regulatory science: Extending the controls governing the quality of public research to private research. Am. J. Law Med. 30(2-4):119-154.

6

Findings and Recommendations

Since its formation in 1970, the US Environmental Protection Agency (EPA) has played a leadership role in developing the broad fields of environmental science and engineering. It has stimulated and supported basic and applied research, developed environmental-education programs, supported regional science initiatives, supported and promoted the development of safer and more cost-effective technologies, provided a firm scientific basis of regulatory decisions, and prepared the agency to address emerging environmental problems. The broad reach of EPA science has also influenced international policies and guided state and local actions. As a result of EPA's scientific leadership, both the nation and the world have made great progress in addressing environmental challenges and improving environmental quality over the last 40 years.

As a regulatory agency, EPA applies much of its resources to implementing complex regulatory statutes that have been established by Congress. That regulatory mission can engender controversy and place strains on the conduct of EPA's scientific work in ways that do not affect most other government science agencies, such as the National Institutes of Health and the National Science Foundation (NSF). Amid this inherent tension, EPA generally, and the Office of Research and Development (ORD) specifically, strive to meet the following objectives in their research:

- Support the needs of the agency's present regulatory mandates and timetables.
- Identify and lay the intellectual foundations that will allow the agency to address current environmental challenges and challenges that it will face over the course of the next several decades.
- Determine the main environmental problems on the US environmental-research landscape.
- Sustain and continually rejuvenate a diverse inhouse scientific research staff—with the necessary laboratories and field capabilities—to support the agency in its present and future missions and in its active collaboration with other agencies.

- Strike a balance between inhouse and extramural research investment. The latter can often bring new ideas and methods to the agency, stimulate a flow of new people into it, and support the continued health of environmental research in the nation.

In the present climate of tight federal budgets, EPA faces the challenge of how to set priorities and achieve as many of these research objectives as it can within a limited budget that, in some cases, is shrinking in real terms.

The committee has examined the agency's capacity to obtain and apply the best new scientific and technologic tools to meet current and future challenges. For 4 decades, EPA has been a national and world leader in addressing the scientific and engineering challenges of protecting the environment and human health. The agency's multidisciplinary science workforce of 6,000 is bolstered by strong ties to academic research institutions and science advisers representing many sectors of the scientific community. A highly competitive fellowship program also provides a pipeline for future environmental science and engineering leaders and enables the agency to attract graduates who have state-of-the-art training.

Thus, the foundation of EPA science is strong. However, the agency needs to successfully address numerous present and future challenges if it is to maintain science leadership and meet its expanding mandates. There is a pressing need to groom tomorrow's leaders and prepare for the retirement of large numbers of senior scientists (some of whom have been with the agency since it was created in 1970). As this report has underscored, there is an increased recognition of the need for cross-disciplinary training and the expansion of the science base to strengthen capacity in social and information sciences. In addition, EPA will continue to need leadership in the traditional core subjects, including, but not limited to, statistics, chemistry, economics, environmental engineering, ecology, toxicology, epidemiology, exposure science, and risk assessment. EPA's future success will depend on its capacity to address long-standing environmental problems, to recognize and respond to emerging challenges, to develop solutions, and to meet the scientific needs of policy-makers.

Figure 6-1 presents the committee's approach for addressing science for EPA's future. The following sections elaborate on the issues described above and bring together the principal findings and recommendations detailed throughout the report. In assessing the scientific opportunities and needs that the agency faces, the committee did not consider it appropriate to prioritize where EPA should invest its limited resources. Such an exercise will require detailed internal EPA deliberations and administrative guidance. Instead, the committee has focused on the statement of task, which asked for an assessment of EPA's capabilities to develop, obtain, and use new science and technologic information to meet persistent, emerging, and future challenges.

Most of the committee's recommendations, which are discussed in Chapters 4 and 5 and summarized in the sections below, are broad and are intended to help EPA enhance its ability to address environmental problems and their solu-

Findings and Recommendations 189

tions from a systems perspective and through strengthened leadership, communication, internal expertise, and internal and external collaboration. The mechanism or mechanisms through which EPA chooses to address the recommendations will depend on its funding, its priorities, and what environmental science and engineering areas it wants to focus its efforts on in the future. EPA already addresses some aspects of the committee's recommendations to some degree. It is the committee's aim that this report will help the agency to choose where to enhance its ability to integrate its current science and to use new tools and technologies to address its mission challenges.

SYSTEMS THINKING

It is important for EPA to try to balance its capacity and resources to address complex environmental challenges, to address potential favorable and unfavorable health and environmental effects, and to apply emerging scientific information, tools, techniques, and technologies. Approaching problems from a systems perspective will allow EPA to meet those challenges and make the maximum continuing use of new scientific tools. The committee has suggested ways in which the agency can integrate systems-thinking techniques into a 21st century framework for science to inform decisions (see Figure 6-1). That framework will help EPA to stay at the leading edge of science by encouraging it to produce science that is anticipatory, innovative, long-term, and collaborative; to evaluate and apply emerging tools for data acquisition, modeling, and knowledge development; and to develop tools and methods for synthesizing science, characterizing uncertainties, and integrating, tracking, and assessing the outcomes of actions. If effectively implemented, the framework would help to break the silos of the agency and promote collaboration among different media, time scales, and disciplines. In supporting environmental science and engineering for the 21st century, there will need to be a move from using science to characterize risks, to applying science holistically to characterize both problems and solutions at the earliest possible time. ORD's move toward embracing sustainability throughout its research program is a positive move in this direction.

Finding: Environmental problems are increasingly interconnected. EPA can no longer address just one environmental hazard at a time without considering how that problem interacts with, is influenced by, and influences other aspects of the environment.

Recommendation 1: The committee recommends that EPA substantially enhance the integration of systems thinking into its work and enhance its capacity to apply systems thinking to all aspects of how it approaches complex decisions.

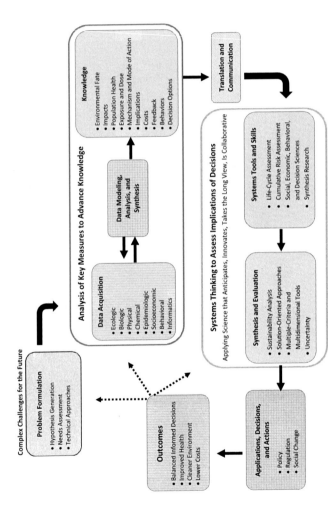

FIGURE 6-1 Framework for enhanced science for environmental protection. The iterative process starts with effective problem formulation, in which policy goals and an orientation toward solutions help to determine scientific needs and the most appropriate methods. Data are acquired as needed and synthesized to generate knowledge about key outcomes. This knowledge is incorporated into an array of systems tools and solutions-oriented synthesis approaches to formulate policies that best improve public health and the environment while taking account of social and economic impacts. Once science-informed actions have been implemented, outcome evaluation can help determine whether refinements to any previous stages are required (see the dotted lines in the figure).

The following paragraphs provide examples of some of strategies that EPA could use to help it set its own priorities and to enhance its use of systems thinking.

Even if formal quantitative life-cycle assessment (LCA) is not feasible, increased use of a life-cycle perspective would help EPA to assess activities, regulatory strategies, and associated environmental consequences. Placing more of a focus on life-cycling thinking would likely include increasing EPA's investment in the development of LCA tools that reflect the most recent knowledge in LCA and risk assessment (both human health and ecologic). In addition, it may be more cost effective for EPA to provide incentives and resources to increase collaboration between LCA practitioners in the agency and those working on related analytic tools (such as risk assessment, exposure modeling, alternatives assessment, and green chemistry). EPA has some internal capacity for LCA, but could benefit from a more systematic use of such an assessment across the agency's mission.

Continuing to invest intramural and extramural resources in cumulative risk assessment and the underlying multistressor data, including coordinated bench science and community-based components, would give EPA a broader and more comprehensive understanding of the complex interactions between chemicals, humans, and the environment. A challenge before the agency is the characterization of cumulative effects using complex, incomplete, or missing data. Even as EPA seeks to improve its understanding of risks, some prevention-based decisions may need to be made in the face of uncertainty.

In EPA's science programs, environmental decisions will only be effective if they consider the social and behavioral contexts in which they will play out. Such decisions can substantially affect societal interests beyond those that are specifically environmental. Tradeoffs among environmental and other societal outcomes need to be anticipated and made explicit if decision-making is to be fully informed and transparent, and predicting economic and societal responses at various points in the decision-making process is necessary to achieve desirable environmental and societal outcomes. For these reasons, developing mechanisms to integrate social, economic, behavioral, and decision sciences would lead to more comprehensive environmental-management decisions. EPA can engage the social, economic, behavioral, and decision sciences as part of a systems-science perspective rather than as consumers and evaluators of others' science. In addition, EPA would benefit from a long-term commitment to advancing research in a number of related fields, including valuation of health and ecosystem benefits.

Research centers that focus on synthesis research have demonstrated the power and cost effectiveness of bringing together multidisciplinary collaborative groups to integrate and analyze data to generate new scientific knowledge. Deliberately introducing synthesis research into EPA's activities would contribute to accelerating its progress in sustainability science. A specific area where knowledge from systems thinking could be applied is in the design of safer chemicals, products, and materials.

ENHANCED SCIENCE LEADERSHIP

The committee evaluated EPA's capabilities and the needs that the agency will face given both large and complex future environmental challenges and the necessity of identifying, evaluating, and implementing a large number of new scientific tools in its science and decision-making. Based on that evaluation, it identified a need to substantially strengthen its science leadership. There has been progress toward agency-wide science integration with the establishment of the Office of the Science Advisor, and further progress might be made with the shift of the science advisor position from within ORD to the Office of the Administrator in early 2012; however, the Office of the Science Advisor may need further authority from the administrator or additional staff resources to continue to improve the integration and coordination of science across programs and regions throughout the agency. When the committee speaks of *enhancing* science leadership, it is not just referring to the strengthened capacity of someone in a higher-level position within EPA to whom the administrator has provided independence, authority, and resources, but also the internal support at all levels in the agency (including scientists, analysts, directors, and deputy and assistant administrators) to ensure that the highest-quality science is developed, evaluated, and applied systematically throughout the agency's programs.

In the committee's analysis of the strengths and limitations of an enhanced agency-wide leadership position, it has concluded that successful implementation of the systems-based application of emerging tools and technologies to meet persistent and future challenges cannot be achieved under the current structure. Success will require leadership throughout the agency, in the programs and regions as well as in ORD. There will need to be clear lines of authority and responsibility, and regional administrators, program assistant administrators, and staff members at all levels will need to be held accountable for ensuring scientific quality and the integration of individual science activities into broader efforts across the agency.

Finding: The need for improvement in the oversight, coordination, and management of agency-wide science has been documented in studies by the National Research Council, the General Accountability Office, and the agency's own Science Advisory Board as a serious shortcoming and it remains an obstacle at EPA. The committee's own analysis of challenges and opportunities for the agency indicates the need for integration of systems thinking, and the need for enhanced leadership at all levels is even stronger than it has been in the past.

Recommendation 2: The committee recommends that the EPA administrator continue to identify ways to substantially enhance the responsibilities of a person in an *agency-wide* science leadership position. That person should hold a senior position, which could be that of a deputy administrator for science, a chief scientist, or possibly a substantially strengthened version of the current science advisor position. He or she should have sufficient au-

thority and staff resources to improve the integration and coordination of science across the agency. **If this enhanced leadership position is to be successful, strengthened leadership is needed throughout the agency and the improved use of science at EPA will need to be carried out by staff at all levels.**

Whatever administrative arrangement is adopted, the following are suggestions of the types of responsibilities that the committee thinks should be associated with this position:

- Chairing and assuring that the work of the Science and Technology Policy Council is comprehensive and effective.
- Promotion of systems thinking and systems-oriented tools to address complex challenges ahead and the integration of this approach into every aspect of agency science and engineering (as described in Chapter 4).
- Working to ensure that the scientific and technical staff throughout the agency (including program, regional, and research offices) have the expertise necessary to perform their duties whether in support of the agency's research or in support of its role as a regulatory and policy decision-maker.
- Assuring that the agency has in place a system for quality assurance and quality control of its scientific and technical work (including a system for consistent high-quality peer review).
- Assuring that the best available scientific and technical information is being used to carry out the agency's mission.
- Working to coordinate research and analytic efforts within and outside the agency to ensure that the best information is used in the most efficient manner.
- Encouraging and supporting interoffice and interagency science collaboration in order to solve problems and develop good solutions.

If the occupant of the position is to be successful, he or she will require sufficient staff and resources to act on behalf of the EPA administrator to implement a coordinated budget and strategic planning process for the regional, program, and research offices to ensure that appropriate scientific and technical expertise and capabilities are available and used. The person in this position would also oversee the policies and procedures that relate to the operation of the agency's federal advisory committees. **The committee specifically recommends that the person in this position and his or her staff create, implement, and periodically update an integrated,** *agency-wide* **multiyear plan for science, its use, and associated research needs.** Such a plan would bring together ORD, program, and regional science initiatives while being cognizant of the flexibility that is imparted through bottom-up initiatives undertaken in ORD, the program offices, and the regions.

The strengthening of science leadership is not without its challenges. For example, whether or not the position is held by a political appointee could affect the ability of the person in the position to be effective throughout the agency, especially with the other political appointees who head the programs that rely on science (and supervise many of the agency's scientists). There is also the possibility that new procedures established from the central administration could serve to discourage innovation in science if not carefully applied. To a certain extent, the recent EPA decision to re-establish the position of science advisor as a non-political position distinct from ORD (as had been the case in earlier EPA administrations) will provide a test of how to overcome some of these challenges. However, the revised role of the current science advisor does not fully implement the committee's recommendation unless that person is empowered with the tools and support described above. Even with the full support of the administrator and senior staff, the effort will fail if the need to improve the use of science in EPA is not accepted by staff at all levels.

STRENGTHENING CAPACITY

Assessing and obtaining the proper scientific expertise within the agency is necessary to address complex environmental problems facing the nation and to create and implement solutions. That includes having the expertise to take advantage of new technologies that will improve the science basis of regulatory decision-making at the national, state, and local levels. It also includes having broad interdisciplinary expertise and engaging in collaboration to more effectively evaluate system-level impacts and sustainable solutions. In order to be prepared to address a wide array of environmental and health challenges and their complex interactions, EPA will need to continue to ensure that it has expertise in critical fields. In some cases, the agency will need to advance scientific understanding through inhouse research; in others, it will need to assimilate and influence scientific efforts that are undertaken elsewhere. However, even as the agency moves to increase the breadth and depth of its skills in new disciplines, and especially in light of an aging work force, continued support is needed to ensure that basic scientific disciplines are strongly represented. In order to have the capacity to address future environmental challenges, the agency will need to have enough internal expertise to identify and collaborate with the expertise of all of its stakeholders so that it can ask the right questions; determine what existing tools and strategies can be applied to answer those questions; determine the needs for new tools and strategies; develop, apply, and refine the new tools and strategies; and use the science to make recommendations based on hazards, exposures, and monitoring.

Finding: EPA has been a leader in environmental science and technology both nationally and internationally. If it is to retain that leadership in the coming decades, it must maintain its expertise in traditional scientific disciplines while enhancing the breadth and depth of its skills in new disciplines.

Recommendation 3: The committee recommends that the agency strengthen its scientific capacity by (a) continuing to cultivate knowledge and expertise within the agency generally, (b) hiring more behavioral and decision scientists, and (c) engaging mechanisms to draw on scientific research and expertise from outside of the agency.

Within EPA

Addressing the environmental challenges of today and the future requires forward-thinking and resourceful scientists and engineers. One of the keys to recruiting and retaining high-quality scientists who can help the agency to maintain its leadership role is for the agency to foster an environment where scientists and engineers have opportunities to work on interesting, challenging problems, interact synergistically with colleagues, have an impact, and earn recognition for their work. Furthermore, if the agency is going to address the problems of today and the future from a systems perspective, its scientists and engineers need to be able to optimize resources, create and benefit from scientific exchange zones, and lead innovation through transdisciplinary collaborations.

Finding: Expertise in traditional scientific disciplines—including, but not limited to, statistics, chemistry, economics, environmental engineering, ecology, toxicology, epidemiology, exposure science, and risk assessment—are essential for addressing the challenges of today and the future. The case of statistics is one example where the agency is facing significant retirements and needs to have, if anything, enhanced expertise. EPA is currently attuned to these needs, but staffing high-quality scientists in these areas of expertise who can embrace problems by drawing from information across disciplines will require continued attention if EPA is to maintain its leadership role in environmental science and technology.

Recommendation 3a: EPA should continue to cultivate a scientific workforce across the agency (including ORD, program offices, and regions) that can take on transdisciplinary challenges.

Some options that EPA might explore to fulfill the recommendation above include:

- Build a stronger mentoring and leadership development program that supports young researchers and fosters a culture of systems-thinking research.
- Recruit young scientists who have expertise and interest in scientific concepts and tools relevant to systems thinking and its supporting analytic tools.

- Promote rotations through its laboratories and through the laboratories of other federal agencies and scientific organizations as valuable training experiences for new scientists in the areas of environmental health, science, and engineering.
- Expand opportunities for internal networking, including opportunities for scientists and engineers to work between programs and offices.
- Encourage scientists and engineers to work in interdisciplinary teams and in new ways to provide expertise where it is needed in a timely fashion.
- Implement programs to help scientists and engineers to acquire new skills and expertise throughout their careers, including educational opportunities, sabbaticals, and other kinds of leave, and laboratory rotations.
- Provide opportunities for agency scientists to interact with colleagues in other agencies, in universities, in nonprofit organizations, and in the private sector; such opportunities could include workshops, roundtables, participation in traditional research conferences, and long-term exchanges with or as visiting scientists.
- Promote the visibility and recognition of scientific excellence across its divisions, programs, and locations by enhancing and highlighting its featured research and awards programs.
- Assess its current policies for retaining and hiring civil service employees. The agency must be nimble and must be able to easily hire or reassign employees to make sure it has specific expertise to understand emerging challenges and make use of new tools, technologies, and approaches in the appropriate offices, regions, and laboratories at the appropriate time.

Economic, social, behavioral, and decision sciences can make important contributions to improving environmental policy decisions within the emerging integrated systems-based approach to environmental management. They can also make contributions to supporting innovative strategies for achieving environmental goals efficiently, equitably, and cooperatively. Behavioral and decision sciences are particularly essential in dealing with such issues as "framing effects", cognitive heuristics, risk communication, and the design and assessment of the likely effectiveness of alternative regulatory strategies.

Finding: EPA's economic, social, behavioral, and decision science staff consists almost entirely of economists. The agency is without strong expertise in social, behavioral, and decision sciences, though it does support some research in these areas through outside grants, collaborations, and procurement.

Recommendation 3b: The committee recommends that EPA add staff who have training in behavioral and decision sciences and find ways to enhance the existing staff capabilities in these fields.

Options that EPA might explore to fulfill that recommendation include:

- Recruit several new staff who have earned advanced degrees in empirically based behavioral and decision sciences. The new staff would need to have strong communication skills and would need to work closely with economists, natural scientists, and engineers in the agency to help to make regulatory and other agency policies that promote environmentally protective behaviors that are more realistic. Their knowledge would assist the agency by helping it to make more informed choices when seeking outside contractors and advisers and to create stronger collaboration with academics in related fields. The committee suggests that the new staff be located within the National Center for Environmental Economics (NCEE). The reason for that suggestion is that NCEE currently staffs the largest number of social scientists within the agency. The large interest in behavioral and decision sciences that exists now in economics broadly, as exhibited by the fields of behavioral and neuroeconomics, will contribute to making NCEE a productive location. More importantly, behavioral economics is an essential source of new insight in environmental economics research pertaining to the benefits of environmental protection and the design of incentives for environmental management. Co-locating behavioral scientists within NCEE will increase the capacity of economics staff to participate in the advances in environmental economics emerging from the integration of behavioral economics.
- Provide mechanisms for cross-disciplinary training of staff in core disciplines that are relevant to behavioral and decision sciences. The committee acknowledges that the number of staff in EPA who have advanced training in these fields is likely to remain modest even with a concerted recruitment effort, and it is important for staff scientists who work in adjacent disciplines to have enough familiarity to know what questions to ask (and whom to ask).
- Develop improved mechanisms for integrating economic, social, behavioral, and decision science into the development of science to support environmental-management decisions.

Outside of EPA

EPA would be well-advised to continue to take advantage of such mechanisms as extramural funding to access the expertise that it needs. One example is the Science to Achieve Results (STAR) program, which is sponsored by EPA's National Center for Environmental Research to support transdisciplinary and interdisciplinary relationships through interactive and collaborative projects. It can also access experts through collaborations. Specifically, it could reestablish the collaborative research program between ORD and the NSF Decision, Risk, and Management Sciences program. That type of collaboration would allow EPA to harness the expertise that it needs to make informed judgments in behavioral and social sciences.

Most of the agency's science needs will probably continue to be met by research and collaboration performed through existing means. However, EPA also has the potential to acquire more information through collaboration with the public. For example, the explosion of new Internet-based, wireless, and miniaturized sensing technologies provides an unprecedented opportunity to involve the public in research and in meeting data-collection needs in ways that were not possible in the past. The emergence of secure enterprise social networks also provides a host of opportunities for EPA to greatly enhance internal and external collaboration. There is potential for the collection of environmental information and the sorting and analysis of complex data to be accomplished through citizen science, crowdsourcing, and similar techniques. EPA will need to continue to follow new and emerging technologies closely and make anticipatory decisions for adoption where its mission can be addressed in a cost-effective way.

Even if resources were not a major constraint, EPA would still need the expertise to be able to harness the science, data, information, tools, techniques, equipment, and expertise available from research being done in other organizations domestically and internationally. As resources dedicated to research become more limited, tracking, gathering, and using such knowledge becomes even more essential.

Finding: Research on environmental issues is not confined to EPA. In the United States, it is spread across a number of federal agencies, national laboratories, and universities and other public-sector and private-sector facilities. There are also strong programs of environmental research in the public and private sectors in many other nations.

Recommendation 3c: The committee recommends that EPA improve its ability to track systematically, to influence, and in some cases to engage in collaboration with research being done by others in the United States and internationally.

The committee suggests the following mechanisms for approaching the recommendation above:

- Identify knowledge that can inform and support the agency's current regulatory agenda.
- Institute strategies to connect that knowledge to those in the agency who most need it to carry out the agency's mission.
- Inform other federal and nonfederal research programs about the science base that the agency currently needs or believes that it will need to execute its mission.
- Seek early identification of new and emerging environmental problems with which the agency may have to deal.

SCIENCE, TOOLS, AND TECHNOLOGIES TO ADDRESS CURRENT AND FUTURE CHALLENGES

As with all of science and engineering, the fields of environmental science and technology continue to evolve. Tools and methods are becoming more powerful and sophisticated. In Chapter 3, the committee identified some examples of tools and technologies that have helped and will continue to help EPA to address challenges that are relevant to its mission. As mentioned at the beginning of this chapter, the committee was not asked to and did not attempt to prioritize specific tools and technologies that EPA should invest in for the future. Those decisions will need to be made by EPA based on factors such as where it would like to develop its inhouse expertise in the future, where it would prefer to collaborate to gain the expertise it needs, and where it would like to leverage or incentivize outside expertise. Some specific areas the committee identified where EPA may want to consider maintaining or enhancing its expertise on in the future include:

- Extend collaborations with remote-sensing scientists.
- Find ways to engage in broader, deeper, and sustained support for long-term monitoring.
- Continue to promote methodologic development and application to rapid and predictive monitoring.
- Develop a quantitative microbial risk-assessment framework that incorporates alternative indicators, using genomic approaches, microbial source tracking, and pathogen monitoring.
- Collaborate with other agencies (for example, the National Institute of Environmental Health Sciences Exposure Biology program; the NSF Environmental, Health, and Safety Risks of Nanomaterials program; the US Centers for Disease Control and Prevention; and the European Commission's Exposure Initiative) to build a greater capacity for exposure science.
- Improve exposure assessment for environmental-epidemiology studies.
- Continue modeling efforts to advance understanding of sources and environmental processes that contribute to particulate matter loadings and consequent health and environmental effects.
- Improve understanding of interactions between climate change and air quality, with a focus on relatively short-lived greenhouse agents, such as ozone, black carbon, and other constituents of particulate matter.
- Develop processes and procedures for effective public communication of the potential public health and environmental risks associated with the increasing number of chemicals.
- Improve understanding of the value and limitations of "-omic" technologies and approaches for environmental and human health risk assessment.
- Continue validation of high-throughput in vitro assays for the screening of new chemicals for potentially hazardous properties while continuing to recognize the limitations and strengths of current toxicity-testing approaches.

Regardless of the specific tools and technologies EPA intends to invest its resources on in the future, it must at least have knowledge of new technologies and tools that are emerging in the areas of environmental science and engineering. EPA's efforts to anticipate science needs and emerging tools to meet these needs cannot succeed in a vacuum. As it focuses on organizing and catalyzing its internal efforts better, it will need to continue to look outside itself—to other agencies, states, other countries, academe, and the private sector—to identify relevant scientific advances and opportunities where collaboration that relies on others' efforts can be the best (sometimes the only) means of making progress in protecting health and the environment.

Finding: Although EPA has periodically attempted to scan for and anticipate new scientific, technology, and policy developments, these efforts have not been systematic and sustained. The establishment of deliberate and systematic processes for anticipating human health and ecosystem challenges and new scientific and technical opportunities would allow EPA to stay at the leading edge of emerging science.

Recommendation 4: The committee recommends that EPA engage in a deliberate and systematic "scanning" capability involving staff from ORD, other program offices, and the regions. Such a dedicated and sustained "futures network" (as EPA called groups with a similar function in the past), with time and modest resources, would be able to interact with other federal agencies, academe, and industry to identify emerging issues and bring the newest scientific approaches into EPA.

IMPROVED MANAGEMENT AND USE OF LARGE DATASETS

Without good data that show the state of the environment, how it is evolving, and how it is affecting people and ecosystems, it is difficult to do an effective, science-based job of environmental protection. EPA is gathering and will continue to gather large amounts of data from a diverse array of sources and will need to deposit such data into data management systems that are both secure and accessible. EPA will need to have the capacity to systematically access, harvest, manage, and integrate data from diverse sources, in different media, across geographic and disciplinary boundaries, and of heterogeneous forms and scales. This capacity will depend on EPA maintaining and possibly increasing its current information-technology capabilities that support state-of-the art data acquisition, storage, and management. Capacity will also depend on having enough senior statisticians in the agency to analyze, model, and support the synthesis of data. EPA will need to continue to promote and engage in the development of informatics techniques for seamless data integration and synthesis and robust model development. As EPA continues to strengthen its informatics infrastructure, including data-warehousing and data-mining, it remains important to pay attention to new analytic and statistical methods, the building blocks of informatics and backbones of data-mining; to address emerging modeling issues; and to bridge methodologic gaps.

Many of the issues being addressed by EPA are in the context of environmental factors whose effects are best characterized in terms of changing exposures, accumulating amounts of materials, and changing health and environmental conditions. Given the high levels of spatial and temporal variability of those factors, it is often critical to have and maintain long-term records of multiple parameters. Making data and samples accessible to future researchers is central to ensuring that the understanding of environmental phenomena continues to grow and evolve with the science. It is also important to develop sample archives where materials are appropriately stored and to have good metadata for analysis or reanalysis at a later date.

Long-term monitoring is essential for tracking changes in ecosystems and populations to identify, at the earliest stage, emerging changes and challenges. Without long-term data, it is difficult to know whether current variations fall within the normal range of variation or are truly unprecedented. It is also essential for knowing whether EPA's management interventions are having their intended effect. Monitoring is a fundamental component of hypothesis-testing. All management interventions are based on explicit or implicit hypotheses that justify them and explain why they should yield the desired results. A hypothesis may focus on physical and biologic processes or on expected human behavioral responses. If it is made explicit and monitoring is designed specifically to test it, both the value of the monitoring and the details of its design will be clarified, and the importance of the monitoring will be evident.

Finding: It is difficult to understand the overall state of the environment unless one knows what it has been in the past, and how it is changing over time. Typically this can only be achieved by examining high-quality time series of key indicators of environmental quality and performance. Currently at EPA, there are few long-term monitoring programs, let alone programs that are systematic and rigorous.

Recommendation 5: The committee recommends that EPA invest substantial effort to generate broader, deeper, and sustained support for long-term monitoring of key indicators of environmental quality and performance.

INNOVATION

To understand future environmental health problems and provide solutions, EPA will depend on innovations across different media (air, land, and water). EPA has an important role in addressing capacity and opportunities for innovation by providing information, technical assistance, platforms for information exchange, demonstration activities, and economic incentives and disincentives. It can play a role not only in promoting innovation in the agency but stimulating innovation by others. The agency also has the opportunity to leverage resources to support innovation. The committee does not recommend that EPA attempt to develop all such solutions itself. Rather, it would be more cost effective to partner and engage with others to support innovation. That can be supported through EPA's Small Business Innovation Research program or an award, such as the Presidential Green Chemistry Awards, which would nudge the entrepreneurial community to address problems of direct interest to the agency. EPA has taken a global leadership role by supporting efforts that focus on innovative solutions-oriented science, including the pollution prevention program, Design for the Environment, and the green chemistry and engineering program. They demonstrate the potential for innovative approaches to advance and use scientific knowledge to protect health and the environment through the redesign of chemicals, materials, and products. They also demonstrate the role that EPA can play in driving decisions by providing high-quality scientific information.

Finding: EPA has recognized that innovation in environmental science, technology, and regulatory strategies will be essential if it is to continue to perform its mission in a robust and cost-effective manner. However, to date, the agency's approach has been modest in scale and insufficiently systematic.

Recommendation 6: The committee recommends that EPA develop a more systematic strategy to support innovation in science, technology, and practice.

In doing this, the agency would be well-advised to work on identifying much more clearly the "signals" that it is or is not sending and to refine them as needed. Clearly identifying signals could be accomplished by seeking to identify the key desired outcomes of EPA's regulatory programs and communicate the desired outcomes clearly to the private and public sectors. The committee has identified several ways in which EPA could address this recommendation.

- Establish and periodically update an agency-wide innovation strategy that outlines key desired outcomes, processes for supporting innovation, and opportunities for collaboration. Such a strategy would identify incentives, disincentives, and opportunities in program offices to advance innovation. It would highlight collaborative needs, education, and training for staff to support innovation.
- Identify and implement cross-agency efforts to integrate innovative activities in different parts of the agency to achieve more substantial long-term innovation. One immediate example of such integration that is only beginning to occur is bringing the work on green chemistry from the Design for the Environment program together with the innovative work on high-throughput screening

in the ToxCast program to apply innovative toxicity testing tools to the design of green chemicals.

- Explicitly examine the effects of new regulatory and nonregulatory programs on innovation while ascertaining environmental and economic effects. This "innovation impact assessment" could, in part, inform the economic evaluation as a structure that encourages technologic innovation that may lead to long-term cost reductions. The assessment could also function as a stand-alone activity to evaluate how regulations could encourage or discourage innovation in a number of activities and sectors. It could help to identify what research and technical support and incentives are necessary to encourage innovation that reduces environmental and health effects while stimulating economic benefits.

STRENGTHENING SCIENCE IN A TIME OF TIGHT BUDGETS

This report has stressed the importance of sustaining and strengthening EPA's present programs of scientific research, applications, and data collection while identifying and pursuing a wide array of new scientific opportunities and challenges. Both are needed to address the complexity of modern problems and both are essential to the agency if it is to continue to provide scientific leadership and high-quality science-based regulation in the years to come.

Specific recommendations related to agency budgets are outside the scope of this study, but the committee feels compelled to note, as did the report *Science Advisory Board Comments on the President's Requested FY2013 Research Budget* (EPA SAB 2012b), that since 2004, the budget for ORD has declined 28.5% in real-dollar terms (gross domestic product–indexed dollars). The reductions have been even greater in a number of specific fields, such as ecosystem research and pollution prevention.

Finding: If EPA is to provide scientific leadership and high-quality science-based regulation in the coming decades, it will need adequate resources to do so. Some of this committee's recommendations, if followed, will allow EPA to address its scientific needs with greater efficiency. But the agency cannot continue to provide leadership, pursue many new needs and opportunities, and lay the foundation for ensuring future health and environmental safety unless the long-term budgetary trend is reversed.

Recommendation 7: The committee recommends EPA create a process to set priorities for improving the quality of its scientific endeavors over the coming decades. This process should recognize the inevitably limited resources while clearly articulating the level of resources required for the agency to continue to ensure the future health and safety of humans and ecosystems.

Appendix A

Statement of Task of the Committee on Science for EPA's Future

A committee of the National Research Council will assess the overall capabilities of the US Environmental Protection Agency (EPA) to develop, obtain, and use the best available new scientific and technological information and tools that will be needed to meet persistent, emerging, and future mission challenges and opportunities across the agency's research and regulatory programs. These challenges and opportunities will include those posed or provided by new scientific knowledge and techniques, new and persistent environmental problems, changes in human activities and interactions, changes in public expectations, new risk-assessment and risk-management paradigms, new models for decision making, and new agency mission requirements. Special consideration will be given to potentially increasing emphasis on trans-disciplinary approaches, systems-based problem solving, scientific and technological innovation, and greater involvement of communities and other stakeholders. The committee will identify and assess transitional options to strengthen the agency's capability to pursue the scientific information and tools that will be needed to meet these challenges and opportunities.

In performing its task, the committee may consider topics such as the following:

- Key factors expected to stimulate major changes in the biophysical and societal environments, research, risk assessment, risk management, and regulatory decision-making.
- Computational, analytic, and anticipatory strategies for strengthening the agency's capabilities to obtain and interpret scientific information to address such changes.
- New methods and bioinformatics tools to support private-sector efforts to create new chemicals and engineering approaches to developing materials,

Appendix A

products, and services that are sustainable and safer for public health and the environment.

- Organizational collaborations, within EPA's Office of Research and Development (ORD) and among EPA offices, other agencies, and other domestic and foreign institutions that could facilitate EPA's ability to anticipate, identify, and respond to new environmental challenges.
- New informatics approaches to collecting, storing, and sharing data; new techniques of measurement, computation, modeling, monitoring, and analysis; and new methods of synthesizing and integrating information across disciplines.
- New methods to measure the costs and benefits of environmental regulation and to anticipate future risk, the perception of that risk (especially before it is well understood), and the response to that risk.
- Improvements to, or further development of, decision-support tools to assist in evaluation of regulatory alternatives, taking into account relevant regulatory decision-making goals and relevant physical, chemical, biological, engineering, and social sciences.
- Approaches to more effectively deal with the inherent tensions among research, risk assessment, and regulatory timeframes.
- Scientific tools and analytic frameworks, including systems-based, trans-disciplinary, and community-based approaches, to address future regulatory challenges, including examples of potential applications of these tools.
- EPA's scientific capabilities (from both a financial and human resource perspective) to successfully deal with the future.
- Other sources of scientific information external to the agency.

Appendix B

Biographical Information on the Committee on Science for EPA's Future

Jerald L. Schnoor (*Chair*) is the Allen S. Henry Chair in Engineering, professor of civil and environmental engineering, professor of occupational and environmental health, codirector of the Center for Global and Regional Environmental Research, and faculty research engineer of the Iowa Institute of Hydraulic Research – Hydroscience and Engineering. Dr. Schnoor's interests are in water-quality modeling, environmental chemistry, and climate change. His present research includes phytoremediation of groundwater contamination and hazardous wastes, water observatory networks, global change, and sustainability. Dr. Schnoor is editor-in-chief of *Environmental Science and Technology*. He is a member of the US Environmental Protection Agency Science Advisory Board, a member of the National Academy of Engineering, and a member of the National Institute of Environmental Health Science National Advisory Environmental Health Sciences Council. He has served on several previous National Research Council committees, most recently as a member of the Committee on Economic and Environmental Impacts of Increasing Biofuels Production. Dr. Schnoor earned a PhD in civil (environmental health) engineering from the University of Texas.

Tina Bahadori resigned from the committee on March 26, 2012, when she was appointed the national program director for chemical safety for sustainability in the US Environmental Protection Agency (EPA) Office of Research and Development. While she served as a member of the committee, she was the managing director for the American Chemistry Council's Long-Range Research Initiative (LRI) program. In that position, she was responsible for the direction of the LRI, which sponsors an independent research program that advances the science of risk assessment for the health and ecological effects of chemicals to support decision-making by government, industry, and the public. Before joining the American Chemistry Council, she was the manager of Air Quality Health Integrated Programs at the Electric Power Research Institute. Dr. Bahadori is the immediate past president of the International Society of Exposure Science and an associate editor of the *Journal of Exposure Science* and *Environmental Epidemiology*. She served as a member of the Board of Scientific Counselors of the US Centers for Disease Control and Prevention (CDC) National Center for Environmental Health (NCEH)/Agency for Toxic Substances and Disease Registry (ATSDR), as a member of the CDC NCEH/ATSDR National Conversation on Public Health and Chemical Exposure Leadership Council, as a peer reviewer for the EPA grants and programs, and as a member of the Chemical Exposure Working Group on the National Children's Study. She has also served on several National Research Council committees, most recently as a member of the Committee to Develop a Research Strategy for Environmental, Health, and Safety Aspects of Engineered Nanomaterials, the Committee on Human and Environmental Exposure Science in the 21st Century, and the Board on Environmental Studies and Toxicology. Dr. Bahadori earned an ScD in environmental science and engineering from the Harvard School of Public Health.

Eric J. Beckman is the George M. Bevier Professor of Engineering in the University of Pittsburgh Department of Chemical and Petroleum Engineering. He is also codirector of the Mascaro Center for Sustainable Innovation, a center in the school of engineering that focuses on the design of sustainable communities. Dr. Beckman's research interests involve the design of green chemical products and molecular design of biomedical devices. Dr. Beckman was honored as the 1992 recipient of the National Science Foundation Young Investigator Award and the 2002 Academic Presidential Green Chemistry Challenge Award. He earned a PhD in polymer science from the University of Massachusetts.

Thomas A. Burke is associate dean for public-health practice and professor of health policy and management at the Johns Hopkins University Bloomberg School of Public Health. He holds joint appointments in the Department of Environmental Health Sciences and the School of Medicine's Department of Oncology. Dr. Burke is also director of the Johns Hopkins Risk Sciences and Public Policy Institute. His research interests include environmental epidemiology and surveillance, evaluation of population exposures to environmental pollutants, assessment and communication of environmental risks, and application of epidemiology and health risk assessment to public policy. Before joining Johns Hopkins University, Dr. Burke was deputy commissioner of health for New Jersey and director of science and research for the New Jersey Department of Environmental Protection. In New Jersey, he directed initiatives that influenced the development of national programs, such as Superfund, the Safe Drinking Water Act, and the Toxics Release Inventory. Dr. Burke was the inaugural chair of the advisory board to the director of the US Centers for Disease Control and Prevention National Center for Environmental Health and is currently a member of the US Environmental Protection Agency Science Advisory Board. He has served on several National Research Council committees, most recently as chair of the

Committee on Improving Risk Analysis Approaches Used by the US EPA and the Committee to Review EPA's Title 42 Hiring Authority for Highly Qualified Scientists and Engineers. He was also chair of the Committee on Human Biomonitoring for Environmental Toxicants and the Committee on Toxicants and Pathogens in Biosolids Applied to Land. In 2003, he was designated a lifetime national associate of the National Academies. Dr. Burke received his PhD in epidemiology from the University of Pennsylvania.

Frank W. Davis is director of the National Center for Ecological Analysis and Synthesis and a professor at the Bren School of Environmental Science and Management of the University of California, Santa Barbara, where he teaches ecology and conservation planning. His current research focuses on the landscape ecology of California rangelands, ecologic implications of modern climate change, and conservation planning for renewable-energy development. Dr. Davis is a fellow of the American Association for the Advancement of Science, a fellow in the Aldo Leopold Leadership Program, a Google Science Communication Fellow, and a trustee of the Nature Conservancy of California. He has served on several National Research Council committees and is currently a member of the Board on Environmental Studies and Toxicology. Dr. Davis earned a PhD in geography and environmental engineering from Johns Hopkins University.

David L. Eaton is a professor of environmental and occupational health sciences and interim vice provost for research at the University of Washington (UW). He also serves as the director of the National Institute of Environmental Health Sciences Center for Ecogenetics and Environmental Health at UW. He has held several other UW positions, including director of the toxicology program and associate chairman in the Department of Environmental Health and associate dean for research in the School of Public Health. Dr. Eaton maintains an active research and teaching program that is focused on the molecular basis of environmental causes of cancer and how human genetic differences in biotransformation enzymes may increase or decrease individual susceptibility to chemicals found in the environment. He has published over 150 scientific articles and book chapters in toxicology and risk assessment. Nationally, he has served on the board of directors and as treasurer of the American Board of Toxicology, as secretary and later as president of the Society of Toxicology, as a member of the board of directors and as vice-president of the Toxicology Education Foundation, and as a member of the board of trustees of the Academy of Toxicological Sciences. Dr. Eaton is a member of the Institute of Medicine and has served on several National Academies committees, most recently as a member of the Institute of Medicine Committee on Breast Cancer and the Environment: The Scientific Evidence, Research Methodology, and Future Directions. He is an elected fellow of the American Association for the Advancement of Science and of the Academy of Toxicological Sciences. Dr. Eaton earned a PhD in pharmacology from the University of Kansas Medical Center.

Paul Gilman is senior vice president and chief sustainability officer of Covanta Energy. Previously, he served as director of the Oak Ridge Center for Advanced Studies and as assistant administrator for the Office of Research and Development in the US Environmental Protection Agency. He also worked in the Office of Management and Budget, where he had oversight responsibilities for the US Department of Energy (DOE) and all other science agencies. In DOE, he advised the secretary of energy on scientific and technical matters. From 1993 to 1998, Dr. Gilman was the executive director of the Commission on Life Sciences and the Board on Agriculture and Natural Resources of the National Research Council. He has served on numerous National Research Council committees and is currently a member of the Committee on Human and Environmental Exposure Science in the 21st Century. Dr. Gilman received his PhD in ecology and evolutionary biology from Johns Hopkins University.

Daniel S. Greenbaum is president and chief executive officer of the Health Effects Institute (HEI), an independent research institute funded jointly by government and industry. In this role, he leads HEI's efforts to provide public and private decision-makers with high-quality, impartial, relevant, and credible science on the health effects of air pollution to inform air-quality decisions in the developed and developing world. Mr. Greenbaum has focused HEI's efforts on providing timely and critical research and reanalysis on particulate matter, air toxics, diesel exhaust, and alternative technologies and fuels. Before joining HEI, he served as commissioner of the Massachusetts Department of Environmental Protection. Mr. Greenbaum has chaired the US Environmental Protection Agency (EPA) Blue Ribbon Panel on Oxygenates in Gasoline and EPA's Clean Diesel Independent Review Panel, and he is a member of the board of directors of the Environmental Law Institute. He has also served on several National Research Council committees, most recently the Committee on Health, Environmental, and Other External Costs and Benefits of Energy Production and Consumption and the Committee on Estimating Mortality Risk Reduction Benefits from Decreasing Tropospheric Ozone Exposure. Mr. Greenbaum earned an MS in city planning from the Massachusetts Institute of Technology.

Steven P. Hamburg is chief scientist at the Environmental Defense Fund. He is an ecosystem ecologist specializing in the impacts of disturbance on forest structure and function. He has served as an adviser to both corporations and nongovernment organizations on ecologic and climate-change mitigation issues. Previously, he spent 16 years as a tenured member of the Brown University faculty and was founding director of the Global Environment Program of the Watson Institute for International Studies. Dr. Hamburg is the co-chair of the Royal Society's Solar Radiation Management Governance Initiative and a member of the US Department of Agriculture Advisory Committee on Research, Economics, Extension and Education. He has been the recipient of several awards, including recognition by the Intergovernmental Panel on Climate Change for contributing to its being awarded the 2007 Nobel Peace Prize. Dr. Hamburg earned a PhD in forest ecology from Yale University.

James E. Hutchison is the Lokey-Harrington Professor of Chemistry at the University of Oregon. He is the founding director of the Oregon Nanoscience and Microtechnologies Institute's Safer Nanomaterials and Nanomanufacturing Initiative, a virtual center that unites 30 principal investigators in the Northwest around the goals of designing greener nanomaterials and nanomanufacturing. Dr. Hutchison's research focuses on molecular-level design and synthesis of functional surface coatings and nanomaterials for a wide array of applications in which the design of new processes and materials draws heavily on the principles of green chemistry. He has received several awards and honors including the Alfred P. Sloan Research Fellowship and the National Science Foundation Career Award. He was a member of the National Research Council Committee on Grand Challenges for Sustainability in the Chemistry Industry and he is currently a member of the Committee to Develop a Research Strategy for Environmental, Health, and Safety Aspects of Engineered Nanomaterials. Dr. Hutchison received a PhD in organic chemistry from Stanford University.

Jonathan I. Levy is a professor of environmental health at the Boston University School of Public Health. Dr. Levy's research centers on air pollution exposure assessment and health risk assessment, with a focus on urban environments and issues of heterogeneity and equity. Major research topics include evaluating spatial patterns of air pollution in complex urban terrain, developing methods to quantify the magnitude and distribution of health benefits associated with emissions controls for motor vehicles and power plants, using systems science approaches to evaluate the influence of indoor environmental exposures on pediatric asthma in low-income housing, and developing methods for community-based cumulative risk assessment that includes chemical and non-chemical stressors. Dr. Levy was the recipient of the Health Effects Institute Walter A. Rosenblith New Investigator Award in 2005. He has been a member of several National Research Council committees, including the Committee on Health Impact Assessment and the Committee on Improving Risk Analysis Approaches Used by the US Environmental Protection Agency. Dr. Levy earned his ScD in environmental science and risk management from the Harvard School of Public Health.

David E. Liddle joined US Venture Partners as a general partner in 2000 after retiring as president and chief executive officer of Interval Research Corporation, a laboratory and new-business incubator in Silicon Valley, California. He has spent his career in developing technologies for interaction and communication between people and computers in activities spanning research, development, management, and entrepreneurship. Before cofounding Interval, Dr. Liddle founded Metaphor Computer Systems in 1982 and served as its president and chief executive officer. The company was acquired by IBM in 1991, and Dr.

Liddle was named vice president of new-systems business development. Previously, he held various research and development positions in Xerox Corporation's Palo Alto Research Center. Dr. Liddle has served as director of numerous public and private companies and as chair of the board of trustees of the Santa Fe Institute. He has served on the Defense Advanced Research Projects Agency Information Science and Technology Committee and has participated in several National Research Council committees, including as chair of the Computer Science and Telecommunications Board, member of the Committee on Innovation in Information Technology, and chair of the Committee to Study Wireless Technology Prospects and Policy. He has been named a senior fellow of the Royal College of Art and elected as a director of the New York Times Company. Dr. Liddle earned a PhD in electrical engineering from the University of Toledo.

Jana B. Milford is a professor of mechanical engineering at the University of Colorado. She previously served as a senior staff member with the Environmental Defense Fund. Her research addresses technical, legal, and policy aspects of air pollution. Her primary technical focus is modeling the chemistry and transport of ozone, secondary organic aerosols, and other photochemical air pollutants. Her research includes application of formal sensitivity and uncertainty analysis and optimization techniques to chemistry and transport models and use of these models in making decisions. She is also interested in application and evaluation of statistical and mass-balance techniques for identifying sources of air pollution on the basis of chemically speciated measurements, including outdoor, indoor, and personal exposure measurements. She has served on several National Research Council committees, including the Committee on Air Quality Management in the United States, and is currently a member of the Board on Environmental Studies and Toxicology. Dr. Milford obtained a PhD from the Department of Engineering and Public Policy of Carnegie Mellon University and a JD from the University of Colorado School of Law.

M. Granger Morgan is a professor and head of the Department of Engineering and Public Policy and University and Lord Chair Professor in Engineering at Carnegie Mellon University. In addition, he holds academic appointments in the Department of Electrical and Computer Engineering and in the H. John Heinz III College. His research addresses problems in science, technology, and public policy with a particular focus on energy, environmental systems, climate change, and risk analysis. Much of his work has involved the development and demonstration of methods of characterizing and treating uncertainty in quantitative policy analysis. At Carnegie Mellon, Dr. Morgan directs the National Science Foundation Climate Decision Making Center and is codirector of the Carnegie Mellon Electricity Industry Center. He serves as chair of the Scientific and Technical Council for the International Risk Governance Council. In the recent past, he served as chair of the US Environmental Protection Agency Science Advisory Board and as chair of the Electric Power Research Institute Advisory Council. He is a fellow of the American Association for the Advancement of

Science, the Institute of Electrical and Electronics Engineers, and the Society for Risk Analysis. He is a member of the National Academy of Sciences and serves as a member of the National Academies Division Committee on Engineering and Physical Sciences, the Report Review Committee, the Aeronautics Research and Technology Roundtable, the Keck Futures Initiative Ecosystem Services Steering Committee, and the Planning Committee on Fostering Partnerships and Linkages in Sustainability Science and Innovation—A Symposium. Dr. Morgan earned a PhD in applied physics from the University of California, San Diego.

Ana Navas-Acien is an associate professor in the Department of Environmental Health Sciences of the Johns Hopkins Bloomberg School of Public Health. She is a physician–epidemiologist with a specialty in preventive medicine and public health and a long-term interest in the health consequences of widespread environmental exposures. Her research, based on an epidemiologic approach, investigates chronic health effects of arsenic, selenium, lead, cadmium, and other trace metals. Dr. Navas-Acien has served as an expert witness to the Baltimore City Council and served as a member of the 2010 National Toxicology Program Workshop on the Role of Environmental Chemicals in the Development of Diabetes and Obesity. She earned an MD from the University of Granada School of Medicine in Spain and a PhD in epidemiology from the Johns Hopkins School of Public Health.

Gordon H. Orians is a professor emeritus of biology at the University of Washington (UW). He was a professor at UW from 1960 to 1995 and was director of the UW Institute for Environmental Studies from 1976 to 1986. Most of Dr. Orians's research has focused on behavioral ecology of birds and has dealt primarily with problems of habitat selection, mate selection and mating systems, selection of prey and foraging patches, and the relationship between ecology and social organization. Recently, his research has focused on environmental aesthetics and the evolutionary roots of strong emotional responses to components of the environment, such as landscapes, flowers, sunsets, and sounds. Dr. Orians has served on the Science Advisory Board of the US Environmental Protection Agency and on boards of such environmental organizations as the World Wildlife Fund and the Nature Conservancy. He has also served on many National Academies committees, including the Committee on Independent Scientific Review of Everglades Restoration Progress, the Committee on Cumulative Environmental Effects of Alaskan North Slope Oil and Gas Activities, and the Board on Environmental Studies and Toxicology. He is a member of the National Academy of Sciences and the American Academy of Arts and Sciences. Dr. Orians earned a PhD in zoology from the University of California, Berkeley.

Joan B. Rose serves as the Homer Nowlin Chair in Water Research at Michigan State University, the codirector of the Center for Advancing Microbial Risk Assessment, and the director of the Center for Water Sciences. She is an international expert in water microbiology, water quality, and public-health safety and has over 300 publications. Dr. Rose's work has examined new molecular methods for detecting waterborne pathogens and zoonotic agents and source-tracking techniques. She has been involved in the study of water supplies, water used for food production, coastal environments, and drinking-water treatment, wastewater treatment, reclaimed water, and water reuse. She has been instrumental in advancing quantitative microbial risk assessment. Dr. Rose was awarded the Athelie Richardson Irvine Clarke Prize from the National Water Research Institute for contributions to water science and technology in 2001 and the International Water Association (IWA) Women in Water award in 2008 and is currently a member of the Strategic Council of the IWA. She had served as chair of the US Environmental Protection Agency Science Advisory Board Drinking Water Committee. She is a member of the National Academy of Engineering and has served on several National Academies committees, most recently the Planning Committee for Water Challenges for Public Health Needs Domestically and Internationally: A Workshop, the Committee on Sustainable Underground Storage of Recoverable Water, and the Panel on Human Health and Security. Dr. Rose earned a PhD in microbiology from the University of Arizona.

James S. Shortle is Distinguished Professor of Agricultural and Environmental Economics and director of the Environment and Natural Resources Institute of Pennsylvania State University. His research focuses on markets and incentives for ecosystem services with a goal of advancing theory and practice. He is also interested in the use of integrated assessment for environmental decision-making to improve capacity to predict, manage, and adapt to environmental change. Dr. Shortle has served on the editorial boards of *Environment and Development Economics* and *European Review of Agricultural Economics.* He has served as a member and secretary of the National Technical Advisory Committee of the US Department of Energy National Initiative on Global Environmental Change, as a member of the US Environmental Protection Agency (EPA) Science Advisory Board (SAB) Panel on the Second Generation Model, and as a member of the National Research Council Committee on Water Quality in the Pittsburgh Region, and he is currently a member of the EPA SAB Environmental Economics Advisory Committee. Dr. Shortle earned a PhD in economics from Iowa State University.

Joel A. Tickner is an associate professor in the Department of Community Health and Sustainability of the University of Massachusetts Lowell. He is interested in the development of innovative scientific methods and policies to implement a precautionary and preventive approach to decision-making under uncertainty while advancing assessment and adoption of safer substitutes to chemicals and products of concern. His teaching and research interests include regulatory science and policy, risk assessment, pollution prevention, cleaner production, and environmental health. Dr. Tickner has served on several advisory boards and as an expert reviewer, most recently for the California Green Chemistry Initiative, the US Environmental Protection Agency National Pollution Prevention and Toxics Advisory Committee, and the First National Precautionary Principle Conference Advisory Committee. He is the recipient of several honors and awards, including the University of Massachusetts President's Award for Public Service, the National Pollution Prevention Roundtable Champion Award, and the North American Hazardous Waste Managers Policy Leader Award. Dr. Tickner earned an ScD in cleaner production and pollution prevention from the University of Massachusetts Lowell.

Anthony D. Williams is founder and chief executive of Anthony D. Williams Consulting. He is an author, speaker, and consultant who helps organizations to harness the power of collaborative innovation in business, government, and society. He is a coauthor of *Wikinomics: How Mass Collaboration Changes Everything* and the followup book *MacroWikinomics: Rebooting the Business and the World*. Mr. Williams is currently a visiting fellow at the Munk School of Global Affairs of the University of Toronto and a senior fellow for innovation at the Lisbon Council in Brussels. Among other appointments, he is an adviser to GovLoop, the world's largest social network for government innovators, and a founding fellow of the OpenForum Academy, a global research initiative focused on understanding the effects of open standards and open sources on business and society. As a senior fellow at nGenera Insight, Mr. Anthony founded and led the world's definitive investigation into the impact of Web 2.0 and wikinomics on the future of governance and democracy. He has advised Fortune 500 firms and international institutions, including the World Bank. Mr. Williams earned an MSc in research in political science from the London School of Economics.

Yiliang Zhu is a professor in the Department of Epidemiology and Biostatistics of the University of South Florida College of Public Health. He is also director of the college's Center for Collaborative Research. His current research is focused on biostatistical methods for spatiotemporal data, exposure to environmental contaminants and health consequences, evaluation of health-care systems and outcomes, and quantitative methods in health risk assessment, including physiologically based pharmacokinetic and pharmacodynamic models, dose–response modeling, benchmark-dose methods, and uncertainty quantification. He also conducts research in disease surveillance and health-care access and use in developing countries. Dr. Zhu has served as a member of several National Research Council committees as is currently a member of the Committee on Shipboard Hazard and Defense II (SHAD II). He received his PhD in statistics from the University of Toronto.

Appendix C

The Rapidly Expanding Field of "–Omics" Technologies

Technologic advances in "-omics" technologies—especially in the genomics, proteomics, metabolomics, bioinformatics, and related fields of the molecular sciences (referred to here collectively as panomics)—have transformed the understanding of biologic processes at the molecular level and should eventually allow detailed characterization of molecular pathways that underlie the biologic responses of humans and other organisms to environmental perturbations. The following sections discuss recent advances in –omics technologies and approaches. They also discuss some of the implications of –omics technologies for the US Environmental Protection Agency (EPA), areas in which EPA is at the leading edge of applying the technologies to address environmental problems, and the areas in which EPA could benefit from more extensive engagement.

GENOMICS

Beginning in the late 1990s, the Human Genome Project (DOE 2011) ushered in an unprecedented leap in technologies that allow scientists to discern the fundamental sequences of genes of entire genomes—not only the human genome but a plethora of model organisms, such as plants, microorganisms, invertebrates, vertebrates, and even the long-extinct woolly mammoth (Miller et al. 2008; NHGRI 2012). The ability to derive, quickly and relatively inexpensively, the entire sequence of an organism's genome provides unprecedented opportunities in biologic and ecologic sciences, including the opportunity to understand how environmental factors influence biology at the molecular level.

The Human Genome Project fueled the development of faster and less expensive DNA sequencing. So called first-generation sequencing technologies, originally described by Sanger and Coulson (1975), have served as the primary technology for DNA sequencing for the last several decades, with estimated costs of $3 billion to sequence the human genome (NHGRI 2010; Woollard et al. 2011). Large-scale sequencing projects based on several next-generation sequencing technologies can now be conducted faster and less expensively than was possible with previous generations of technologies. Next-generation sequencing technologies are substantially different from those based on the original Sanger method (Box C-1) and promise remarkable increases in sequencing capabilities.

Next-generation sequencing instruments have made it possible to sequence huge amounts of DNA quickly, thoroughly, and affordably and have opened opportunities to study a wide array of biologic questions, from the metagenomics of water, to characterization of the genetic basis of species differences in response to environmental insults, to human variability in susceptibility to environmentally related diseases. Third-generation sequencing promises to provide full genome sequencing of individuals (humans or other organisms) for less than $1,000 per genome by the end of 2013 (Valigra 2012), and at least one company already offers such services at about $5,000 per genome (Knome 2012).

TRANSCRIPTOMICS

The sequencing of the human genome, and of the genomes of hundreds of other model organisms of great importance for human and environmental health constitutes an enormous step forward in understanding genetic origins of disease, genetic variability, evolutionary biology, and many other subjects of scientific relevance to EPA. However, from a biologic perspective, it is the expression of the genes in specific cells and tissues that ultimately defines an organism and how it responds to its environment. Thus, measuring the extent of gene expression at a given time in a particular cell or tissue is potentially even more informative of biologic mechanisms. The universe of small RNA molecules that are transcribed from DNA and that are present in a cell or tissue at any given time is referred to as the transcriptome. In the last 2 decades, new tools have been developed that allow one to analyze the entire transcriptome in a cell or tissue and to study changes in gene expression that might be created by changes in the environment, such as exposure to a chemical. There are now microarray methods that allow for the analysis of virtually all mRNA molecules that are transcribed from active genes. Typically, these arrays contain hundreds of thousands of unique features that quantitatively identify the amount of a particular mRNA transcript in the sample. Having multiple features that can use the array to look at different parts of a single gene, such as different exons or exon–intron boundaries (potential splice sites), provides a remarkable snapshot of what genes are functioning in a cell at a particular time.

To study complex and common diseases that may be influenced by environmental factors (such as cardiovascular disease and cancer), human studies typically require high-quality DNA from thousands of patients, often from small quantities of tissues or blood. Several common commercial microarrays for RNA applications in studies of this sort have been available for more than a decade and measure the expression of individual genes. However, understanding the human transcriptome is much more complex than simply measuring the complement of mRNAs from the genome because alternative splicing[1] is common and contributes largely to protein and functional diversity in humans and other higher organisms (Xu et al. 2011). Technologies for measuring mRNA transcripts in all their varieties, including alternatively spliced transcripts and copy-number variants, have grown rapidly in the last few years. For example, a new approach called the Glue Grant Human Transcriptome Array completes a comprehensive analysis of the human transcriptome using a 6.9 million–feature oligonucleotide array. The array assesses gene-level and exon-level expression by using high-density tiling of probes that cover a large collection of transcriptome. It can also detect alternative splicing and can analyze noncoding transcripts and common variants (such as single nucleotide polymorphisms) of genes (so called cSNPs) (Xu et al. 2011). This technology was recently used in a multicenter clinical program that produced high-quality reproducible data (Xu et al. 2011). It is an example of the rapid change in technologies in the -omics world and will increasingly provide new approaches to understanding how environmental factors influence the development of common diseases. Such technologies will also have many applications in the fields of microbial genomics, evolutionary biology, and other areas of interest to EPA.

BOX C-1 Comparison of Sanger and Next-Generation Sequencing (NGS)

- The initial preparation of the DNA sample is more labor intensive for NGS than for Sanger, but the amount of sequence data obtained per sample is substantially more.
- The number of sequencing reads from a single instrument per run is of the order of thousands with Sanger, but millions to billions with NGS; for example, a bacterial genome can be sequenced in a single run in days using NGS, versus months using Sanger sequencing.
- Read lengths from Sanger sequencing are up to 900 [base pairs], but in NGS vary from 30 to 500 [base pairs] depending on the platform.
- DNA sequencing costs have been driven down by NGS and base pair per dollar costs show a consistent 19-months doubling time reduction for Sanger sequencing. For NGS, the equivalent figure is approximately 5-months doubling time cost reduction.
- NGS can detect somatic mutations at [less than or equal to] 1%, whereas Sanger sequencing has significantly less sensitivity.
- The greater versatility of NGS is illustrated in generating whole-genome datasets, such as miRNA and ChIP-Seq; Sanger sequencing lacks this capability.

Abbreviations: ChIP-Seq, chromatin immunoprecipitation sequencing; miRNA, micro RNA; NGS, next-generation sequencing. Source: Woollard et al. 2011.

---

[1]Alternative splicing "the process by which individual exons of pre-mRNAs are spliced to produce different isoforms of mRNA transcripts from the same gene" (Xu et al. 2011).

PROTEOMICS

Proteomics is the study of the entire complement of proteins in a cell or tissue—the proteome. The proteome is much more complicated than the genome because the proteome differs from cell to cell and from time to time, whereas the genome of an organism is largely unchanged between cells and over time. Furthermore, most proteins in a cell undergo posttranslational modifications (for example, phosphorylation, glycosylation, methylation, and ubiquination), which can result in several functional forms of the same protein. The proteome is potentially far more informative than the genome with respect to environmental response. Measuring and understanding changes in the proteome after environmental perturbations are therefore increasingly important in many fields of environmental science and engineering. Proteomic technologies and approaches will have an increasingly important role in environmental monitoring and health risk assessment of relevance to EPA. For example, proteome-based biomarkers may be useful in deciphering the associations between pesticide exposure and cancer and will perhaps lead to potential predictive biomarkers of pesticide-induced carcinogenesis (George and Shukla 2011).

Proteomics has been used to explore "a multitude of bacterial processes, ranging from the analysis of environmental communities [and the] identification of virulence factors to the proteome-guided optimization of production strains" (Chao and Hansmeier 2012). Proteomics has become a valuable tool for the global analysis of bacterial physiology and pathogenicity, although many challenges remain, especially in the accurate prediction of phenotypic consequences based on a given proteome composition (Chao and Heinsmeyer 2012). Lemos et al. (2010) have discussed the advantages of and challenges to using proteomics in ecosystems research.

METABOLOMICS

Substantial improvements in instrumentation, especially nuclear magnetic resonance spectroscopy (Serkova and Niemann 2006) and mass spectrometry (Dettmer et al. 2007), provide increasingly sensitive approaches to measuring hundreds or even thousands of small molecules in a cell in a matter of minutes. The new technologies have given rise to a promising new -omics technology referred to as metabolomics—the "systematic study of the unique chemical fingerprints that specific cellular processes leave behind" (Bennett 2005) or, more specifically, the study of their small-molecule metabolite profiles. "In analogy to the genome, which is used as synonym for the entirety of all genetic information, the metabolome represents the entirety of the metabolites within a biological system" (Oldiges et al. 2007). The total number of metabolites in a single cell, tissue, or organism is, of course, highly variable and depends on the biologic system investigated. Hundreds of distinct metabolites have been identified in microorganisms. For example, the *Escherichia coli* database EcoCYC contains over 2,000 metabolite entries (Keseler et al. 2011), and the metabolome of the common baker's yeast, *Saccharomyces cerevisiae,* has about 600 metabolites, the major ones having molecular weight below 300 g/mol (reviewed in Oldiges et al. 2007). It has been projected that plants have more than 200,000 primary and secondary metabolites (Mungur et al. 2005).

Although far less mature than transcriptomics and proteomics, metabolomics offers great promise for the development of early biomarkers of disease (Hollywood et al. 2006) and other uses of relevance to EPA. Because metabolomics in many ways is the final integration of genomics, transcriptomics, and proteomics, it is likely that future developments in this area will become essential for understanding the functions of the genomes of organisms of interest to EPA, ranging from pathogenic bacteria in drinking water to humans. Indeed, EPA scientists are applying metabolomics approaches to aquatic toxicology (Ekman et al. 2011), in vitro assessments for developmental toxicology (Kleinstreuer et al. 2011), and carcinogenic risk assessment (Wilson et al. 2012 in press), to name a few.

EPIGENETICS

As noted by Rothstein et al. (2009), "epigenetics is one of the most scientifically important, and legally and ethically significant, cutting-edge subjects of scientific discovery." Epigenetic changes are the chemical alterations or chemical modifications of DNA that do not involve changes in the nucleotide sequence in the DNA. Those alterations play a critical role in how and when a particular gene is expressed. It is clear that environmental factors, including diet, can influence how epigenetic regulation of gene expression occurs. It is especially important during periods of cell and tissue growth, such as embryonic and fetal development. Epigenetic changes can be triggered by environmental factors. For example, exposure to metals, persistent organic pollutants, and some endocrine disruptors modulate epigenetic markers in mammalian cells and in other environmentally relevant species and have the potential to cause disease (Vandegehuchte and Janssen 2011; Guerrero-Bosagna and Skinner 2012). Some studies have demonstrated that epigenetic changes can sometimes be transferred to later generations, even in the absence of the external factors that induced the epigenetic changes (Skinner 2011).

EPA scientists in the National Health and Environmental Effects Research Laboratory (NHEERL) are aware of the growing importance of epigenetics in environmental health assessment. A seminal review of the application of epigenetic mechanisms to carcinogenic risk assessment was published by NHEERL's scientist Julian Preston (2007). Since then, relatively few publications from NHEERL or other EPA laboratories have addressed epigenetics. A PubMed search identified five publications by EPA scientists in the last 5 years. A recent review by Jardim (2011) discussed the implications of microRNAs (a form of epigenetic regulation of gene expression) for air-pollution research, and Lau et al. (2011) reviewed fetal programming of adult disease (also thought to be an epigenetic phenomenon) and its implications for prenatal care. Hsu et al. (2007)

addressed the implications of epigenetics in the carcinogenic mode of action of nitrobenzene, but only two original research publications that provided experimental data from EPA have directly assessed epigenetic mechanisms. One study (Grace et al. 2011) evaluated the role of maternal influences on epigenetic programming in the in utero development of endocrine signaling in the brain. The second (DeAngelo et al. 2008) provided dose-response data on the development of hepatocellular neoplasia in male mice exposed over a lifetime to trichloroacetic acid, a putative carcinogenic product of trichloroethylene solvent breakdown and a chlorination disinfection byproduct. Although they did not assess epigenetic changes experimentally, they suggested that epigenetic mechanisms might explain the observed tumors inasmuch as the compound was not genotoxic. EPA has not published many original papers on epigenetics, but the EPA grants database lists 36 extramural research grants to universities across the country that are exploring the role of epigenetics in environmental response (EPA 2012). Given the relevance of this emerging field, it is important that EPA scientists and regulators become more active in the accumulation of epigenetic knowledge and its application to human and environmental health risk assessment. Although much remains to be learned about epigenetic phenomena, it is likely to be a critical contributor to many diseases that have both a genetic and environmental component, and will be especially important in understanding how exposures early in life might contribute to disease onset later in life.

BIOINFORMATICS

Rapid advances in biotechnology have resulted in an explosion in -omics data and in information on biochemical and physiologic processes in complex biologic systems. The advent of the internet, new technologies, and high-throughput sequencing has spurred further growth of -omics data and has made it possible to disseminate data globally (Attwood et al. 2011). Since the 1990s, the field of bioinformatics has seen growth in response to the need for the generation, storage, retrieval, processing, analysis, and interpretation of -omics data. It draws on the principles, theories, and methods of the biologic sciences, computer science and engineering, mathematics, and statistics, and it has always been at the core of understanding of biologic processes and disease pathways (Attwood et al. 2011). As the -omics revolution continues, bioinformatics will continue to evolve, and EPA will continue to require inhouse expertise and state-of-the-science capacity in the field.

Analysis of biologic data has evolved from comparisons of various kinds of sequence data (Needleman and Wunsch 1970; Smith and Waterman 1981; Lipman and Pearson 1985) to algorithms that can search various sequence databases. Methods and tools have also been developed for the analysis of sequence, annotation, and expression data in support of a wide variety of applications, such as pattern recognition, protein and RNA structure prediction, micro data analysis (Attwood et al. 2011), and biomarker discovery (Baumgartner et al. 2011; Roy et al. 2011). There is an increasing emphasis on understanding biologic systems

Appendix C 221

through modeling of biologic, physiologic, and biochemical processes (Deville et al. 2003; Ng et al. 2006; Viswanathan et al. 2008;), including gene–gene and protein–protein interactions (Tong et al. 2004; Rual et al. 2005); pathway analysis (Schilling et al. 2000; Wishart 2007; Viswanathan et al. 2008); and network mapping (Lee and Tzou 2009.).

An integrative approach is needed to use different types of databases to identify distinct system components (organized in modules and subnetworks) and to understand their relationships and thereby reduce the complexity of a biologic system as a whole (Lee and Tzou 2009). There are outstanding challenges to the integrative modeling of biologic systems, some of which are summarized in a recent report from the SYSGENET Bioinformatics Working Group (Durrant et al. 2011). Because integrative systems modeling requires synthesizing and harmonizing the analyses of transcriptome, proteome, interactome, metabolome, and phenome data, which are likely to be held in numerous heterogeneous databases, it is critical to improve the interoperability, compatibility, and exchange of software modules that are the foundation of data-processing platforms (such as TIQS and xQTL), database platforms (such as GeneNetwork and XGAP), and data-analysis toolboxes (such as HAPPY and R/QTL). A standard computer language for software development and cloud sourcing would facilitate efficient software dissemination to the bioinformatics community. In addition, further development of public repositories for data models and software source code would promote the use of common data structures and file formats.

To stay at the cutting edge of bioinformatics and take full advantage of its rapid advance, EPA will need a highly skilled bioinformatics workforce that can closely follow the development of trends in bioinformatics tools and software closely. As discussed in Chapter 3, EPA already has a leadership role in bioinformatics as applied to toxicity assessment and is well positioned to contribute to standardization and harmonization processes in the field.

REFERENCES

Attwood, T.K., A. Gisel, N.E. Eriksson, and E. Bongcam-Rudloff. 2011. Concepts, historical milestones and the central place of bioinformatics in modern biology: A European perspective. Chapter 1 in Bioinformatics - Trends and Methodologies, M.A. Mahdavi, ed. InTech-Open Access [online]. Available: http://www.intechopen.com/books/bioinformatics-trends-and-methodologies [accessed Mar. 30, 2012].

Baumgartner, C., M. Osl, M. Netzer, and D. Baumgartner. 2011. Bioinformatic-driven search for metabolic biomarkers in disease. J. Clin. Bioinform. 1:2, doi:10.1186/2043-9113-1-2.

Bennett, D. 2005. Growing pains for metabolomics. The Scientist 19(8):25-28.

Chao, T.C., and N. Hansmeier. 2012. The current state of microbial proteomics: Where we are and where we want to go. Proteomics 12(4-5):638-650.

DeAngelo, A.B., F.B. Daniel, D.M. Wong, and M.H. George. 2008. The induction of hepatocellular neoplasia by trichloroacetic acid administered in the drinking water of the male B6C3F1 mouse. J. Toxicol. Environ. Health A 71(16):1056-1068.

Dettmer. K., P.A. Aronov, and B.D. Hammock. 2007. Mass spectrometry-based metabolomics. Mass Spectrom. Rev. 26(1):51-78.

Deville, Y., D. Gilbert, J. van Helden, and S.J. Wodak. 2003. An overview of data models for the analysis of biochemical pathways. Brief. Bioinform. 4(3):246-259.

DOE (US Department of Energy). 2011. Major Events in the US Human Genome Project and Related Projects. Office of Science, US Department of Energy [online]. Available: http://www.ornl.gov/sci/techresources/Human_Genome/project/timeline.shtml [accessed Mar. 30, 2012].

Durrant, C., M.A. Swertz, R. Alberts, D. Arends, S. Möller, R. Mott, J.C. P. Prins, K.J. van der Velde, R.C. Jansen, and K. Schughart. 2011. Bioinformatics tools and database resources for systems genetics analysis in miceça short review and an evaluation of future needs. Brief. Bioinform. 13(2):135-142.

Ekman, D.R., D.L. Villeneuve, Q. Teng, K.J. Ralston-Hooper, D. Martinovic-Weigelt, M.D. Kahl, K.M. Jensen, E.J. Durhan, E.A. Makynen, G.T. Ankley, and T.W. Collette. 2011. Use of gene expression, biochemical and metabolite profiles to enhance exposure and effects assessment of the model androgen 17β-trenbolone in fish. Environ. Toxicol. Chem. 30(2):319-329.

EPA (US Environmental Protection Agency). 2012. Research Project Search. Extramural Research. US Environmental Protection Agency [online]. Available: http://cfpub.epa.gov/ncer_abstracts/index.cfm/fuseaction/search.welcome [accessed Apr. 4, 2012].

George, J., and Y. Shukla. 2011. Pesticides and cancer: Insights into toxicoproteomic-based findings. J. Proteomics 74(12):2713-2722.

Grace, C.E., S.J. Kim, and J.M. Rogers. 2011. Maternal influences on epigenetic programming of the developing hypothalamic-pituitary-adrenal axis. Birth Defects Res. A Clin. Mol. Teratol. 91(8):797-805.

Guerrero-Bosagna, C., and M.K. Skinner. 2012. Environmentally induced epigenetic transgenerational inheritance of phenotype and disease. Mol. Cell Endocrinol. 354 (1-2):3-8.

Hollywood, K, D.R. Brison, and R. Goodacre. 2006. Metabolomics: Current technologies and future trends. Proteomics 6(17):4716-4723.

Hsu, C.H., T. Stedeford, E. Okochi-Takada, T. Ushijima, H. Noguchi, C. Muro-Cacho, J.W. Holder, and M. Banasik. 2007. Framework analysis for the carcinogenic mode of action of nitrobenzene. J. Environ. Sci. Health C Environ. Carcinog. Ecotoxicol. Rev. 25(2):155-184.

Jardim, M.J. 2011. microRNAs: Implications for air pollution research. Mutat. Res. 717(1-2):38-45.

Jeong, Y., B.F. Sanders, and S.B. Grant. 2006. The information content of high-frequency environmental monitoring data signals pollution events in the coastal ocean. Environ. Sci. Technol. 40(20):6215-6220.

Keseler, I.M., J. Collado-Vides, A. Santos-Zavaleta, M. Peralta-Gil, S. Gama-Castro, L. Muñiz-Rascado, C. Bonavides-Martinez, S. Paley, M. Krummenacker, T. Altman, P. Kaipa, A. Spaulding, J. Pacheco, M. Latendresse, C. Fulcher, M. Sarker, A.G. Shearer, A. Mackie, I. Paulsen, R.P. Gunsalus, and P.D. Karp PD. 2011. EcoCyc: A comprehensive database of *Escherichia coli* biology. Nucleic Acids Res. 39(suppl. 1):D583-D590.

Kleinstreuer, N.C., A.M. Smith, P.R. West, K.R. Conard, B.R. Fontaine, A.M. Weir-Hauptman, J.A. Palmer, T.B. Knudsen, D.J. Dix, E.L. Donley, and G.G. Cezar. 2011. Identifying developmental toxicity pathways for a subset of ToxCast chemicals using human embryonic stem cells and metabolomics. Toxicol. Appl. Pharmacol. 257(1):111-121.

Knome. 2012. Knome, Inc. [online]. Available: http://www.knome.com/ [accessed Apr. 2, 2012].

Lau, C., J.M. Rogers, M. Desai, and M.G. Ross. 2011. Fetal programming of adult disease: Implications for prenatal care. Obstet. Gynecol. 117(4):978-985.

Lee, W.P., and W.S. Tzou. 2009. Computational methods for discovering gene networks from expression data. Brief. Bioinform. 10(4):408-423.

Lemos, M.F., A.M. Soares, A.C. Correia, and A.C. Esteves. 2010. Proteins in ecotoxicology - how, why and why not? Proteomics 10(4):873-887.

Lipman, D.J., and W.R. Pearson. 1985. Rapid and sensitive protein similarity searches. Science 227(4693):1435-1441.

Miller, W., D.I. Drautz, A. Ratan, B. Pusey, J. Qi, A.M. Lesk, L.P. Tomsho, M.D. Packard, F. Zhao, A. Sher, A. Tikhonov, B. Rancy, N. Patterson, K. Lindblad-Toh, E.S. Lander, J.R. Knight, G.P. Irzyk, K.M. Fredrikson, T.T. Harkins, S. Sheridan, T. Pringle, and S.C. Schuster. 2008. Sequencing the nuclear genome of the extinct woolly mammoth. Nature 456(7220):387-390.

Mungur, R., A.D. Glass, D.B. Goodenow, and D.A. Lightfoot. 2005. Metabolite fingerprinting in transgenic *Nicotiana tabacum* altered by the *Escherichia coli* glutamate dehydrogenase gene. J. Biomed. Biotechnol. 2005(2):198-214.

Needleman, S.B., and C.D. Wunsch. 1970. A general method applicable to the search for similarities in the amino acid sequence of two proteins. J. Mol. Biol. 48(3):443-453.

Ng, A., B. Bursteinas, Q. Gao, E. Mollison, and M. Zvelebil. 2006. Resources for integrative systems biology: From data through databases to networks and dynamic system models. Brief. Bioinform. 7(4):318-330.

NHGRI (National Human Genome Research Institute). 2010. The Human Genome Project Completion: Frequently Asked Questions. National Human Genome Research Institute [online]. Available: http://www.genome.gov/11006943 [accessed Apr. 3, 2012].

NHGRI (National Human Genome Research Institute). 2012. Highlights. National Human Genome Research Institute [online]. Available: http://www.genome.gov/ [accessed Apr. 3, 2012].

Oldiges, M., S. Lütz, S. Pflug, K. Schroer, N. Stein, and C. Wiendahl. 2007. Metabolomics: Current state and evolving methodologies and tools. Appl. Microbiol. Biotechnol. 76(3):495-511.

Preston, R.J. 2007. Epigenetic processes and cancer risk assessment. Mutat. Res. 616(1-2):7-10.

Rothstein, M.A., Y. Cai, and G.E. Marchant. 2009. The ghost in our genes: Legal and ethical implications of epigenetics. Health Matrix J. Law Med. 19(1) [online]. Available: http://papers.ssrn.com/sol3/papers.cfm?abstract_id=1140443 [accessed Apr. 6, 2012].

Roy, P., C. Truntzer, D. Maucort-Boulch, T. Jouve, and N. Molinari. 2011. Protein mass spectra data analysis for clinical biomarker discovery: A global review. Brief. Bioinform. 12(2):176-186.

Rual, J.F., K. Venkatesan, T. Hao, T. Hirozane-Kishikawa, A. Dricot, N. Li, G.F. Berriz, F.D. Gibbons, M. Dreze, N. Ayivi-Guedehoussou, N. Klitgord, C. Simon, M. Boxem, S. Milstein, J. Rosenberg, D.S. Goldberg, L.V. Zhang, S.L. Wong, G. Franklin, S. Li, J.S. Albala, J. Lim, C. Fraughton, E. Llamosas, S. Cevik, C. Bex, P. Lamesch, R.S. Sikorski, J. Vandenhaute, H.Y. Zoghbi, A. Smolyar, S. Bosak, R. Sequerra, L. Doucette-Stamm, M.E. Cusick, D.E. Hill, F.P. Roth, and M. Vidal.

2005. Towards a proteome-scale map of the human protein–protein interaction network. Nature 437(7062):1173-1178.

Sanger, F., and A.R. Coulson. 1975. A rapid method for determining sequences in DNA by primed synthesis with DNA polymerase. J. Mol. Biol. 94(3):441-448.

Schilling, C.H., D. Letscher, and B.Ø. Palsson. 2000. Theory for the systemic definition of metabolic pathways and their use in interpreting metabolic function from a pathway-oriented perspective. J. Theor. Biol. 203(3):229-248.

Serkova, N.J., and C.U. Niemann. 2006. Pattern recognition and biomarker validation using quantitative 1H-NMR-based metabolomics. Expert Rev. Mol. Diagn. 6(5):717-731.

Skinner, M.K. 2011. Environmental epigenetic transgenerational inheritance and somatic epigenetic mitotic stability. Epigenetics 6(7):838-842.

Smith, T.F., and M.S. Waterman. 1981. Identification of common molecular subsequences. J. Mol. Biol. 147(1):195-197.

Tong, A.H.Y., G. Lesage, G.D. Bader, H. Ding, H. Xu, X. Xin, J. Young, G.F. Berriz, R.L. Brost, M. Chang, Y. Chen, X. Cheng, G. Chua, H. Friesen, D.S. Goldberg, J. Haynes, C. Humphries, G. He, S. Hussein, L. Ke, N. Krogan, Z. Li, J.N. Levinson, H. Lu, P. Ménard, C. Munyana, A.B. Parsons, O. Ryan, R. Tonikian, T. Roberts, A.M. Sdicu, J. Shapiro, B. Sheikh, B. Suter, S.L. Wong, L.V. Zhang, H. Zhu, C.G. Burd, S. Munro, C. Sander, J. Rine, J. Greenblatt, M. Peter, A. Bretscher, G. Bell, F.P. Roth, G.W. Brown, B. Andrews, H. Bussey, and C. Boone. 2004. Global mapping of the yeast genetic interaction network. Science 303(5659):808-813.

Valigra, L. 2012. Ion Torrent Claims to be First with $1K Genome Sequencer. MHT, January 11, 2012 [online]. Available: http://www.masshightech.com/stories/2012/01/09/daily29-Ion-Torrent-claims-to-be-first-with-1K-genome-sequencer.html [accessed Apr. 3, 2012].

Vandegehuchte, M.B., and C.R. Janssen. 2011. Epigenetics and its implications for ecotoxicology. Ecotoxicology 20(3):607-624.

Viswanathan, G.A., J. Seto, S. Patil, G. Nudelman, and S.C. Sealfon. 2008. Getting started in biological pathway construction and analysis. PLoS Comput. Biol. 4(2):e16.

Wilson, V.S., N. Keshava, S. Hester, D. Segal, W. Chiu, C.M. Thompson, and S.Y. Euling. 2012. Utilizing toxicogenomic data to understand chemical mechanism of action in risk assessment. Toxicol. Appl. Pharmacol in press.

Wishart, D.S. 2007. Current progress in computational metabolomics. Brief. Bioinform. 8(5):279-293.

Woollard, P.M., N.A. Mehta, J.J. Vamathevan, S. Van Horn, B.K. Bonde, and D.J. Dow. 2011. The application of next-generation sequencing technologies to drug discovery and development. Drug Discov. Today 16(11/12):512-519.

Xu, W., J. Seok, M.N. Mindrinos, A.C. Schweitzer, H. Jiang, J. Wilhelmy, T.A. Clark, K. Kapur, Y. Xing, M. Faham, J.D. Storey, L.L. Moldawer, R.V. Maier, R.G. Tompkins, W.H. Wong, R.W. Davis, and W. Xiao. 2011. Human transcriptome array for high-throughput clinical studies. Proc. Natl. Acad. Sci. USA 108(9):3707-3712.

Appendix D

Scientific Computing, Information Technology, and Informatics

INFORMATICS AND INFORMATION TECHNOLOGY

IT and informatics are in rapid transition. Technologic change in the global capacity of computing and telecommunication has been growing exponentially (Hilbert and Lopez 2011). The end of Moore's law[1] of exponential growth in computer hardware power (Robert 2000) will require, for example, mastery of parallel programming to sustain the growth of computing performance and to meet the need for analyzing massive amounts of data until a postsilicon era is realized. The next 10 years will see a massive rebuilding of IT infrastructure everywhere.

The economics of IT are also changing profoundly, largely under the favorable pressure of consumer applications. Enormous increases in data bandwidth (especially wireless) have made possible a wide array of mobile endpoints for applications, and this trend will continue. The inability of traditional relational databases to scale to handle the rapid growth in unstructured, semiconsistent, real-time data on which decisions often need to be made based in the commercial world has led to the emergence of such tools as Map Reduce, Hadoop, and other next-generation data environments (NoSQL 2012), which are discussed above. Virtualization is steadily eliminating the concept of a dedicated server in a fixed location, and cloud computing is transforming the economics of IT. Social networking, already a major consumer phenomenon, has now entered the scientific workplace and can be used for heightened collaboration, as discussed above.

All of the emerging changes will require a more responsive and flexible approach to the opportunities afforded by global informatics and lead to a sys-

---

[1]Moore's law is a rule of thumb in the history of computing hardware whereby the number of transistors that can be placed inexpensively on an integrated circuit doubles about every 2 years (Moore 1965).

tems perspective of data instead of a focus on one locale, one experiment, or one medium at a time. Those are the directions that IT and informatics are taking. The challenge will lie in understanding how to harness information for EPA's science needs for the future and understanding the role of advanced computer science and informatics in EPA.

High-Performance Computing

EPA's National Computer Center in Research Triangle Park, North Carolina, houses many of the agency's computing resources, including the supercomputing resources used by the Environmental Modeling and Visualization Laboratory and resources for such major applications as computational toxicology, exposure research, and risk assessment. Those resources are traditional high-performance computing machines, the products of a shrinking and struggling industry segment. The future of high-performance computing machines will look entirely different, and it is important that EPA adjust to the change to remain at the leading edge of the field.

Parallel Programming

Central processing units (CPUs) can no longer be made to run faster, so progress requires putting multiple CPUs, or "cores", on each chip to operate concurrently. That, in turn, requires a decomposition of applications into independent components that can run in parallel. An important opportunity afforded by the effort to create highly parallel programs is that they can also be exported to external networks of underused processing for the few jobs that require massive resources. The existing tools for that style of programming are poor, and the skill is seldom taught. Fortunately, EPA has had experience in this regard in its supercomputing projects, but it will need to expand its overall skills inventory greatly to continue to take advantage of parallel and emerging techniques in computing as Moore's law is repealed.

Cloud Computing

Cloud computing will redefine the economics of computation for the next 20 years. A cloud-computing server typically provides services to its clients in three ways: complete applications (software as a service, or SaaS); a platform for clients to build on (PaaS); or a raw infrastructure of processors, storage, and networks (IaaS). Clouds generally are classified as public (provided commercially), private (to one or more organizations), or hybrid (public with a secure connection to private). Services can be scaled up or down in capacity and performance instantly; the client is charged for the amount of time, storage, CPUs, and bandwidth, moment by moment. Even organizations with extreme needs for computation, storage, and bandwidth and high volatility of demand over the short term have been able to transition from their own data centers to the cloud with excellent results (Cockcroft 2011). EPA has recognized the opportunity presented by cloud computing and has begun to embark on a process of transition for many services to a private EPA cloud (Lee and Eason 2010).

Throughout EPA, and especially in the regions and the technical offices, applications and databases are the responsibility of regions and offices, but the Office of Technology Operations and Planning (in the Office of Environmental Information) provides the infrastructure, platform, and support from datacenters in Research Triangle Park, North Carolina; Arlington, Virginia; Chicago, Illinois; and Denver, Colorado. Thus, it is natural for EPA scientific computing to move to PaaS and IaaS cloud operation, and it has begun to do so. Done carefully, this will also permit some applications to be moved to the public cloud as economics requires. Given the trajectory of costs and budgets, that is inevitable, and it is important that EPA continue on this path, ensuring that new science applications are designed for private cloud implementation and for later portability to the public cloud.

Wireless Networks

Dramatic improvement in the performance of data transmission in both wide-area and local wireless networks is driving enormous growth in mobile devices and applications. With many government agencies upgrading infrastructure under pressure to use more effectively the underused radiofrequency spectrum over which they have control, that growth will continue for the foreseeable future. Combined with new-generation real-time sensors, the wireless network has a profound effect on collection of and access to environmental information but it also changes expectations about the user experience. Furthermore, designing for mobile devices has different constraints and freedoms from building Web applications for a desktop environment. The techniques will be important as EPA works to engage and gain support from the public. It will be important for EPA to master the skills of spectrum-sharing and efficient use of bandwidth.

DATA MANAGEMENT

With centralized data centers, strong data-quality standards, and highly organized exchanges, EPA is executing well in IT and has adapted to changing technology while continuing to support its original charter to protect the environment and human health. However, a persistent challenge in such fields such as computational toxicology is the integration of available data from many sources. In particular, many investigators who generate large datasets may not have the knowledge and experience in informatics to integrate and interpret the data successfully. In the future, adopting a systems-thinking approach will result in a mixture of data from a variety of sources, including the atmosphere, soil, water, and foods; data will be related to genetics and health outcomes; and they will range from highly unstructured to highly structured data. These factors will require even more multidisciplinary collaboration among agency scientists.

Warehousing and Mining

As increasingly large amounts of data continue to be generated through designated systems—such as environmental monitoring, biomarker and other exposure surveillance data, disease surveillance, and designed epidemiologic and experimental studies—or streamed from community crowdsourcing, EPA is faced with both an opportunity and a challenge of channeling and integrating data into a massive "data warehouse". Data warehousing is a well-developed concept and a common practice in business (Miller et al. 2009). In EPA, the adaptation of and transition to data warehousing will continue to evolve with good protocols, such as EPA's Envirofacts Warehouse (Pang 2009; Egeghy et al. 2012) and the Aggregated Computational Toxicology Resource (Egeghy et al. 2012; Judson et al. 2012). In the future, data in EPA's warehouse will come from diverse sources, from multiple media, and across geographic, physical, and institutional boundaries. Recent efforts to integrate the US Geological Survey's National Water Information System with EPA's Storage and Retrieval System are an example (Beran and Piasecki 2009). To harvest relevant information from massive datasets to support EPA's science and regulatory activities, integration of heterogeneous databases and mining of these massive datasets present some new opportunities. A recent application involving the European Union's Water Resource Management Information System is a case in point (Dzemydienė et al. 2008).

Data-mining has become a standard for analyzing massive, multisource, heterogeneous data on consumer behavior used in business (Ngai et al. 2009). EPA should and can adopt this data analytic paradigm to support its knowledge-discovery process. The paradigm is increasingly important at a time when the discovery of new evidence or a new data model can be bolstered by dynamic mining of large amounts of data, including environmental indicators of air and water, satellite imagery of climate change from representative population databases, health indicators from disease surveillance systems and medical databases, social behavioral patterns, individual lifestyle data, and -omics data and disease pathways. That will require EPA to invest its resources to continue the development of new analytic and computational methods to deal with static datasets (for example, modeling of complex biologic systems and air and water models) and to adapt and develop new data-mining techniques to process, visualize, link, and model the massive amounts of data that are streaming from multiple sources. EPA is making progress in that direction in its Aggregated Computational Toxicology Resource System (Judson et al. 2012). Successful cases have also been reported for ecologic modeling (Stockwell 2006), air-pollution management (Li and Shue 2004), and toxicity screening (Helma et al. 2000; Martin et al. 2009), to name a few.

Large Datasets

Informatics, data warehousing, and data-mining afford EPA powerful tools for maximal use of wealth of information that will continue to be gathered by it, other agencies, and the public on an unprecedented scale. Data analysis and modeling in many cases will be accomplished through informatics techniques, as is already the case in the analysis of -omics data (Ng et al. 2006; Baumgartner et al. 2011; Roy et al. 2011). As EPA moves forward with analyzing and modeling large sets of data, it should keep three points in mind:

- Information generation and information gathering are accelerating exponentially, and EPA will not be able to generate all the data needed to address complex environmental and health problems. It would benefit the agency to continue to develop its capacity to access, harvest, manage, and integrate data from diverse sources and different media and across geographic and disciplinary boundaries rapidly and systematically.
- Links between environmental change, exposure, human behavior, and human health are complex, and seamless integration and dynamic mining of diverse datasets will boost the chance of discovering such links. For example, to derive personal exposure estimates for particulate matter smaller than 2.5 μm in diameter ($PM_{2.5}$), it is necessary to integrate environmental data, human behavioral data, and insight about how $PM_{2.5}$ penetrates various indoor microenvironments. The exposure estimates are then linked to disease-mechanism data and health data. Such an approach is not difficult to appreciate in principle, but its practice hinges on how successfully an informatics approach can be adapted to mine the massive data from diverse systems. EPA has been a leader in air-quality research and associated health effects of exposure to air pollutants, as showcased through its contributions to the Six Cities Study (Dockery et al. 1993) and the National Morbidity, Mortality, and Air Pollution Study (Samet et al. 2000; Dominici et al. 2006), and it is in a strong position to retain its cutting-edge position by adapting informatics approaches to the analysis and modeling of diverse and massive datasets.
- As environmental challenges continue to emerge and evolve, EPA's approach to problem-solving will need to be dynamic and adaptive. Having a cutting-edge capacity of data warehousing, data-mining, bioinformatics, environmental informatics, and health informatics will boost EPA's ability to integrate massive external data in a timely fashion, to adopt new techniques, to borrow scientific and technical expertise from outside the agency, and to be more responsive and anticipatory.

As EPA continues to strengthen its informatics infrastructure, it will be important to pay attention to new analytic and statistical methods to address emerging modeling issues and to bridge methodologic gaps. Several outstanding issues warrant high priority. One challenge is to analyze large amounts of data from diverse sources without having a shared standard for the data collection (Hall et al. 2005). For example, screening and identifying complex chemical mixtures in the natural environment are difficult because there so many possible mixtures and the mixtures change temporally and spatially (Casey et al. 2004). A second example involves conducting gene-screening analysis to differentiate among tens of thousands of genes or single-nucleotide polymorphisms along a hypothesized disease pathway with only a small number of subjects. Overzealous findings of a positive association are a consequence of this high-dimension problem (Rajaraman and Ullman 2011). Mining that type of data could pose serious challenges in validity and utility when the data are from across geographic and disciplinary boundaries and have heterogeneous quality standards. A special danger with huge datasets is a problem of multiple comparisons, which can lead to massive false positive results. Also with such data, there is sometimes a dominance of bias over randomness—increasing the amount of data generally reduces variances, sometimes close to zero, it but does not reduce bias. In fact, it may even increase bias by diverting attention from the basic quality of the data. Another challenge involves the modeling of complex biologic systems (such as pathway models, physiologically based pharmacokinetic and pharmacodynamic models, and hospital admission data). Information from a small number of static datasets is insufficient to support a large number of unknown model parameters. Two approaches are widely used: fixing some parameters at values that have only weak support from external systems (Wang et al. 1997) and tightening the range of variation of the values of the parameters by imposing probabilistic distributions in a Bayesian approach, such as a Monte Carlo Markov chain (Bois 2000). Those methods may give the user an unwarranted sense of truth when there are substantial uncertainties in the true model. As informatics and data-mining become standard, techniques for data analysis will be increasingly hybrid, combining mathematical, computational, graphical, and statistical tools and qualitative methods to conduct data exploration, machine learning, modeling, and decision-making. Developing its inhouse capability will help EPA to adopt and apply the new techniques.

Data Sharing and Distribution

EPA devotes substantial resources to the public sharing of data resources. It also provides support and encouragement to software and application (app) developers for the creation of both institutional and consumer applications for accessing, presenting, and analyzing available environmental data. One example is the Toxics Release Inventory. Others being developed are the EPA Saves Your Skin mobile telephone app, which provides ZIP code–based ultraviolet index information to help the public take action to protect their skin and an air-quality index mobile app, which feeds air-quality information based on ZIP code. The agency has made strides in analytic and simulation activities, as shown in the leadership role that it has played in computational toxicology (see the section "Example of Using Emerging Science to Address Regulatory Issues and Support Decision-Making: ToxCast Program" in Chapter 3).

As information trends move from long-term data to data that are gathered in nearly real time from dispersed geographic sites, there will not be time for a traditional cycle in which the desired information needs to be extracted from the original compilation, reformatted to a specific standard, and finally loaded into an analytic application. It will instead be necessary to literally "send the algorithm to the data" and receive and collect the results centrally. In other words, the complex formulas developed to analyze the data may be used at the site and time of data collection rather than being sent to a central data-processing site for analyses. That approach, first developed by Google in 2004, is named Map Reduce and uses a functional programming model (Dean and Ghemawat 2004). Hadoop, a widely available implementation of Map Reduce, is available in open-source form and from several major vendors. Not only can Hadoop programming parallelize the problem of accessing widely distributed data; it is especially useful for processing unstructured data or combining them with traditional structured data.

REFERENCES

Baumgartner, C., M. Osl, M. Netzer, and D. Baumgartner. 2011. Bioinformatic-driven search for metabolic biomarkers in disease. J. Clin. Bioinform. 1:2, doi:10.1186/2043-9113-1-2.

Beran, B., and M. Piasecki. 2009. Engineering new paths to water data. Comput. Geosci. 35(4):753-760.

Bois, F.Y. 2000. Statistical analysis of Fisher et al. PBPK model of trichloroethylene kinetics. Environ. Health Perspect. 108(suppl. 2):275-282.

Casey, M., C. Gennings, W.H. Carter, V.C. Moser, and J.E. Simmons. 2004. Detecting interaction(s) and assessing the impact of component subsets in a chemical mixture using fixed-ratio mixture ray designs. J. Agr. Biol. Environ. Stat. 9(3):339-361.

Cockcroft, A. 2011. Net Cloud Architecture. Velocity Conference, June 14, 2011 [online]. Available: http://www.slideshare.net/adrianco/netflix-velocity-conference-2011 [accessed Apr. 10, 2012].

Dean, J., and S. Ghemawat. 2004. MapReduce: Simplified data processing on large clusters. Pp. 137-149 in Proceedings of the 6th Symposium on Operating Systems Design and Implementation (OSDI '04), December 5, 2004, San Francisco, CA [online]. Available: http://static.usenix.org/event/osdi04/tech/full_papers/dean/dean.pdf [accessed Mar. 30, 2012].

Dockery, D.W., C.A. Pope, III, X. Xu, J.D. Spengler, J.H. Ware, M.E. Fay, B.G. Ferris, and F.A. Speizer. 1993. An association between air pollution and mortality in six US cities. N. Engl. J. Med. 329(24):1753-1759.

Dominici, F., R.D. Peng, M.L. Bell, L. Pham, A. McDermott, S.L. Zeger, and J.M. Samet. 2006. Fine particulate air pollution and hospital admission for cardiovascular and respiratory diseases. JAMA 295(10):1127-1134.

Dzemydienė, D., S. Maskeliūnas, and K. Jacobsen. 2008. Sustainable management of water resources based on web services and distributed data warehouses. Technol. Econ. Dev. Econ. 14(1):38-50.

Egeghy, P.P., R. Judson, S. Gangwal, S. Mosher, D. Smith, J. Vail, and E.A. Cohen Hubal. 2012. The exposure data landscape for manufactured chemicals. Sci. Total Environ. 414(1):159-166.

Hall, P., J.S. Marron, and A. Neeman. 2005. Geometric representation of high dimension, low sample size data. J. R. Statist. Soc. B 67(3):427-444.

Helma, C., E. Gottmann, and S. Kramer. 2000. Knowledge discovery and data mining in toxicology. Stat. Method. Med. Res. 9(4):329-358.

Hilbert, M., and P. López. The world's technological capacity to store, communicate, and compute information. Science 332(6025):60-65.

Judson, R.S., M.T. Martin, P. Egeghy, S. Gangwal, D.M. Reif, P. Kothiya, M. Wolf, T. Cathey, T. Transue, D. Smith, J. Vail, A. Frame, S. Mosher, E.A. Cohen-Hubal, and A.M. Richard. 2012. Aggregating data for computational toxicology applications: The US Environmental Protection Agency (EPA) Aggregated Computational Toxicology Resource (ACToR) System. Int. J. Mol. Sci. 13(2):1805-1831.

Lee, M., and W. Eason. 2010. The Silver Lining of Cloud Computing. Presentation at Environmental Information Symposium 2010-Enabling Environmental Protection through Transparency and Open Government, May 13, 2010, Philadelphia, PA [online]. Available: http://www.epa.gov/oei/symposium/2010/lee.pdf [accessed Apr. 2, 2012].

Li, S.-T., and L.-Y. Shue. 2004. Data mining to aid policy making in air pollution management. Expert Sys. Appl. 27(3):331-340.

Martin, M.T., R.S. Judson, D.M. Reif, R.J. Kavlock, and D.J. Dix. 2009. Profiling chemicals based on chronic toxicity results from the US EPA ToxRef Database. Environ. Health Perspect. 117(3):392-399.

Miller, F.P., A.F. Vandome, and J. McBrewster, Jr., eds. 2009. Data Warehouse: Extract, Transform, Load, Metadata, Data Integration, Data Mining, Data Warehouse Appliance, Database Management System, Decision Support System. Orlando, FL: Alpha Press.

Moore, G. 1965. Cramming more components onto integrated circuits. Electronics 38(8) [online]. Available: http://www.cs.utexas.edu/~fussell/courses/cs352h/papers/moore.pdf [accessed Apr. 6, 2012].

Ng, A., B. Bursteinas, Q. Gao, E. Mollison, and M. Zvelebil. 2006. Resources for integrative systems biology: From data through databases to networks and dynamic system models. Brief. Bioinform. 7(4):318-330.

Ngai, E.W.T., L. Xiu, and D.C.K. Chau. 2009. Application of data mining techniques in customer relationship management: A literature review and classification. Expert Syst. Appl. 36(2):2592-2602.

NoSQL. 2012. NoSQL Website [online]. Available: http://nosql-database.org/ [accessed Apr. 30, 2012].

Pang, L. 2009. Best practices in data warehousing. Pp. 146-152 in Encyclopedia of Data Warehousing and Mining, 2nd Ed., J. Wang, ed. Hershey, PA: Information Science Reference.

Rajaraman, A., and J.D. Ullman. 2011. Mining of Massive Datasets. New York: Cambridge University Press.

Robert, L.G. 2000. Beyond Moore's law: Internet growth trends. Computer 33(1):117-119.

Roy, P., C. Truntzer, D. Maucort-Boulch, T. Jouve, and N. Molinari. 2011. Protein mass spectra data analysis for clinical biomarker discovery: A global review. Brief. Bioinform. 12(2):176-186.

Samet, J.M., F. Dominici, F.C. Curriero, I. Coursac, and S.L. Zeger. 2000. Fine particulate air pollution and mortality in 20 US cities, 1987–1994. N. Engl. J. Med. 343(24):1742-1749.

Stockwell, D.R.B. 2006. Improving ecological niche models by data mining large environmental datasets for surrogate models. Ecol. Model. 192(1-2):188-196.

Wang, X., M.J. Santostefano, M.V. Evans, V.M. Richardson, J.J. Diliberto, and L.S. Birnbaum. 1997. Determination of parameters responsible for pharmacokinetic behavior of TCDD in female Sprague–Dawley rats. Toxicol. Appl. Pharmacol. 147(1):151-168.